AQA

GCSE biology

Authors

Simon Broadley

Mark Matthews

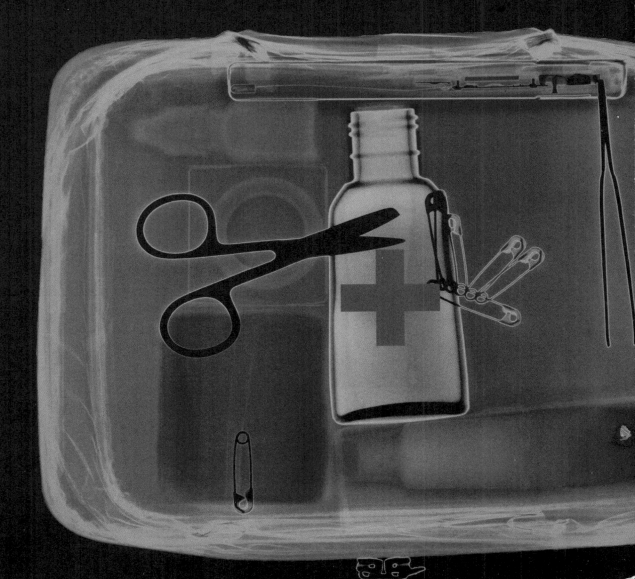

Contents

How to use this book

Welcome to your AQA GCSE Biology revision guide. This book has been specially written by experienced teachers and examiners to match the 2011 specification.

On this page you can see the types of feature you will find in this book. Everything in the book is designed to provide you with the support you need to help you prepare for your examinations and achieve your best.

Specification and student book reference: These show how the pages in the unit match to the exam specification and to your textbook so you can track your progress through the unit as you learn.

Key words: These are the terms you need to understand for your exams.

Exam tip: These hints will help you to think about what may come up in the exam.

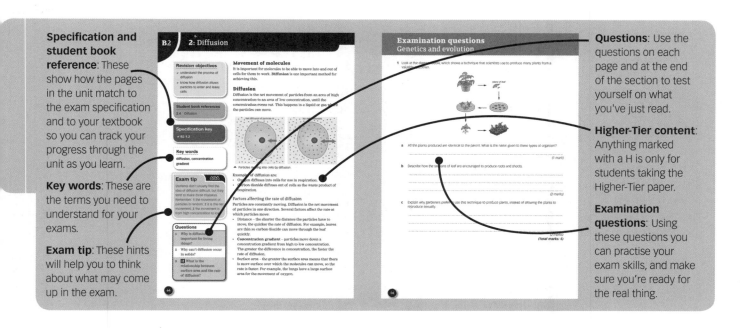

Questions: Use the questions on each page and at the end of the section to test yourself on what you've just read.

Higher-Tier content: Anything marked with a H is only for students taking the Higher-Tier paper.

Examination questions: Using these questions you can practise your exam skills, and make sure you're ready for the real thing.

Upgrade: Upgrade takes you through an exam question in a step-by-step way, showing you why different answers get different grades. Using the tips on this page you can make sure you achieve your best by understanding what each question needs and what an examiner is looking for in your answer.

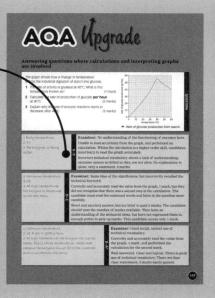

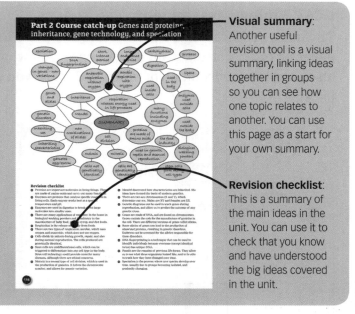

Visual summary: Another useful revision tool is a visual summary, linking ideas together in groups so you can see how one topic relates to another. You can use this page as a start for your own summary.

Revision checklist: This is a summary of the main ideas in the unit. You can use it to check that you know and have understood the big ideas covered in the unit.

Matching your course

The units in this book have been written to match the specification for **AQA GCSE Biology**.

In the diagram below you can see that the units and part units can be used to study either for **GCSE Biology** or as part of **GCSE Science** and **GCSE Additional Science** courses.

	GCSE Biology	GCSE Chemistry	GCSE Physics
GCSE Science	B1 (Part 1)	C1 (Part 1)	P1 (Part 1)
	B1 (Part 2)	C1 (Part 2)	P1 (Part 2)
GCSE Additional Science	B2 (Part 1)	C2 (Part 1)	P2 (Part 1)
	B2 (Part 2)	C2 (Part 2)	P2 (Part 2)
	B3 (Part 1)	C3 (Part 1)	P3 (Part 1)
	B3 (Part 2)	C3 (Part 2)	P3 (Part 2)

GCSE Biology assessment

The units in this book are broken into two parts to match the different types of exam paper on offer. The diagram below shows you what is included in each exam paper. It also shows you how much of your final mark you will be working towards in each paper.

Unit		%	Type	Time	Marks available
Unit 1	B1 (Part 1) / B1 (Part 2)	25%	Written exam	1 hr	60
Unit 2	B2 (Part 1) / B2 (Part 2)	25%	Written exam	1 hr	60
Unit 3	B3 (Part 1) / B3 (Part 2)	25%	Written exam	1 hr	60
Unit 4	Controlled Assessment	25%		1 hr 30 mins + practical	50

When you read the questions in your exam papers you should make sure you know what kind of answer you are being asked for. The list below explains some of the common words you will see used in exam questions. Make sure you know what each word means. Always read the question thoroughly, even if you recognise the word used.

Calculate
Work out your answer by using a calculation. You can use your calculator to help you. You may need to use an equation; check whether one has been provided for you in the paper. The question will say if your working must be shown.

Describe
Write a detailed answer that covers what happens, when it happens, and where it happens. The question will let you know how much of the topic to cover. Talk about facts and characteristics. (Hint: don't confuse with 'Explain')

Explain
You will be asked how or why something happens. Write a detailed answer that covers how and why a thing happens. Talk about mechanisms and reasons. (Hint: don't confuse with 'Describe')

Evaluate
You will be given some facts, data or other information. Write about the data or facts and provide your own conclusion or opinion on them.

Outline
Give only the key facts of the topic. You may need to set out the steps of a procedure or process – make sure you write down the steps in the correct order.

Show
Write down the details, steps or calculations needed to prove an answer that you have been given.

Suggest
Think about what you've learnt in your science lessons and apply it to a new situation or a context. You may not know the answer. Use what you have learnt to suggest sensible answers to the question.

Write down
Give a short answer, without a supporting argument.

Top tips

Always read exam questions carefully, even if you recognise the word used. Look at the information in the question and the number of answer lines to see how much detail the examiner is looking for.

You can use bullet points or a diagram if it helps your answer.

If a number needs units you should include them, unless the units are already given on the answer line.

Revision objectives

- ✔ know what a healthy diet is
- ✔ know that inherited factors can affect health
- ✔ understand what the metabolic rate is, and the factors that affect it
- ✔ know the benefits of regular, frequent exercise

Student book references

1.1 Diet and exercise

1.2 Diet and health

Specification key

- ✔ B1.1.1

A healthy diet

We need to eat food for three reasons:
- for growth and repair
- for energy
- to keep us healthy.

Food is made out of chemicals. Biologists divide the chemicals into five main groups. All foods are a mix of these chemicals.

Food/nutrient	Why you need to eat it	Sources
Carbohydrates	for energy	bread, pasta, rice, sugary food
Fats	for energy	butter, cheese, fried food
Proteins	to build cells and repair tissue	meat, fish, cheese, nuts
Mineral ions and vitamins	needed in small amounts to keep the body healthy	fresh fruit and vegetables

A healthy diet is a **balanced** one, with the right balance of foods and energy to match the body's needs.

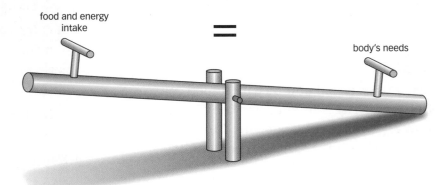

food and energy intake

body's needs

Malnourishment

A person is malnourished if their diet is not balanced. There are basically two types of malnourishment:
- Being underweight – where people do not eat enough, or they **exercise** too much. Their diet has too little food or energy. They lose body mass.
- Being overweight – where people eat too much or exercise too little. Their diet contains too much food or energy for their needs. They gain body mass.

An effect of malnourishment might be that the body lacks some food groups, such as vitamins or minerals. This can lead to **deficiency diseases**, for example, rickets, which is caused by a lack of vitamin D.

Another effect might be an excess of some food groups. Being overweight or eating too much sugar could lead to type 2 diabetes. Too much fat in the diet leads to **obesity** and a high level of cholesterol in the blood. Cholesterol can block blood vessels, leading to heart problems or strokes. Some people have inherited genes that affect their cholesterol levels, increasing their risk of heart disease.

Key words

balanced diet, deficiency disease, obesity, metabolism, metabolic rate, exercise

Metabolic rates

All the cells in your body carry out chemical reactions. This is called **metabolism**. The rate at which all of these reactions occur is called the **metabolic rate**.

Factors that affect the metabolic rate

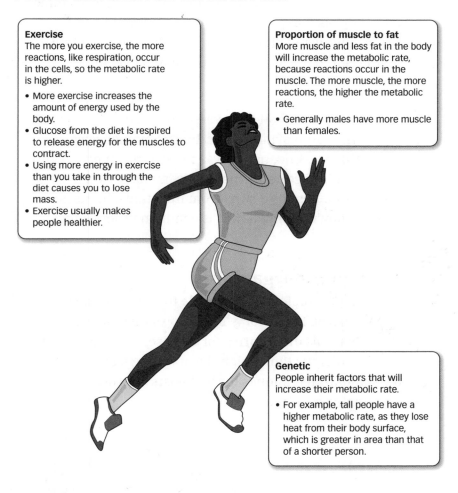

Exercise
The more you exercise, the more reactions, like respiration, occur in the cells, so the metabolic rate is higher.

- More exercise increases the amount of energy used by the body.
- Glucose from the diet is respired to release energy for the muscles to contract.
- Using more energy in exercise than you take in through the diet causes you to lose mass.
- Exercise usually makes people healthier.

Proportion of muscle to fat
More muscle and less fat in the body will increase the metabolic rate, because reactions occur in the muscle. The more muscle, the more reactions, the higher the metabolic rate.

- Generally males have more muscle than females.

Genetic
People inherit factors that will increase their metabolic rate.

- For example, tall people have a higher metabolic rate, as they lose heat from their body surface, which is greater in area than that of a shorter person.

Questions

1 What is a balanced diet?

2 What is your metabolic rate?

3 **H** Explain why too much cholesterol is bad for your health.

Exam tip AQA

Different people have different balanced diets. Think about the lifestyle of the person and what their needs for energy and food might be.

Revision objectives

- ✔ know what a pathogen is
- ✔ know that bacteria and viruses reproduce rapidly inside the body, and produce toxins
- ✔ know the technique for growing microorganisms
- ✔ understand the role of painkillers and antibiotics in treating diseases

Student book references

1.3 Infectious diseases

1.4 Antibiotics and painkillers

Specification key

✔ B1.1.2

Pathogens

A **pathogen** is any **microorganism** that causes an infectious disease. They include some **bacteria** and viruses.

Bacteria

Not all bacteria are pathogens. When bacteria infect our body, they reproduce rapidly, and may produce poisons, called **toxins**, which make us unwell. There are many different types of pathogenic bacteria, and they will cause different diseases. There are even millions of bacteria on our skin.

Viruses

Viruses are much smaller than bacteria. When viruses infect our body, they need to get into our cells. There they will reproduce rapidly and damage our cells, bursting out and causing the cell to release toxins, which make us ill.

Hygiene

Even before biologists had discovered that microorganisms cause disease, a Hungarian doctor, Ignaz Semmelweiss, recognised that washing hands was important. In his hospital he showed that if doctors washed their hands between patients, the numbers of deaths from infectious diseases decreased. Today washing hands is common practice in personal hygiene.

Growing microorganisms

It is important for biologists to be able to grow microorganisms like bacteria in the laboratory. This allows them to test treatments, such as **antibiotics** for disease, or to investigate how effective disinfectants might be at killing bacteria. There are now standard techniques for biologists to grow uncontaminated cultures.

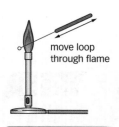

Sterilise the inoculating loop to kill all microorganisms, by passing through a flame.

Dip the loop into a culture of bacteria.

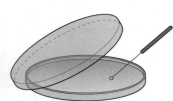

Open a **sterile** Petri dish containing culture media as a gel, and spread the microorganisms.

Incubate at 25 °C in school to reduce the chances of dangerous pathogens growing; in industry higher temperatures are used for more rapid growth.

Seal with tape to stop contamination with microorganisms from the air.

Medicines to treat disease

We can take medicines for two reasons when we have an infection.

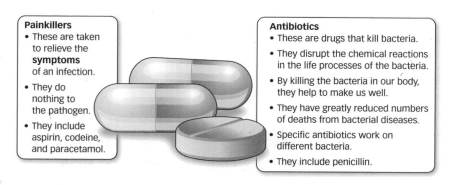

Painkillers
- These are taken to relieve the **symptoms** of an infection.
- They do nothing to the pathogen.
- They include aspirin, codeine, and paracetamol.

Antibiotics
- These are drugs that kill bacteria.
- They disrupt the chemical reactions in the life processes of the bacteria.
- By killing the bacteria in our body, they help to make us well.
- They have greatly reduced numbers of deaths from bacterial diseases.
- Specific antibiotics work on different bacteria.
- They include penicillin.

Key words

microorganism, pathogen, bacteria, virus, toxin, symptoms, painkiller, antibiotic, resistance, natural selection, sterile

Since viruses don't carry out chemical reactions, antibiotics don't kill them. Viruses live inside our cells, so any antiviral drug often harms our own cells.

Resistance to antibiotics

By chance some bacteria can develop a mutation that gives them **resistance** to an antibiotic. They will survive by **natural selection** and form a resistant strain of the bacteria. These infect other people. Gradually the whole population of the bacteria will become resistant to the antibiotic. As antibiotic-resistant strains of bacteria have developed, scientists have had to make new antibiotics. Overuse or inappropriate use of antibiotics increases the chances of resistant bacteria developing.

There is now a problem strain of bacteria, MRSA, which is resistant to most antibiotics. It spreads rapidly as there is no effective treatment.

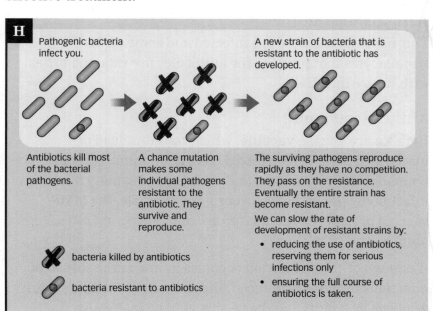

H

Pathogenic bacteria infect you.

A new strain of bacteria that is resistant to the antibiotic has developed.

Antibiotics kill most of the bacterial pathogens.

A chance mutation makes some individual pathogens resistant to the antibiotic. They survive and reproduce.

The surviving pathogens reproduce rapidly as they have no competition. They pass on the resistance. Eventually the entire strain has become resistant.

We can slow the rate of development of resistant strains by:
- reducing the use of antibiotics, reserving them for serious infections only
- ensuring the full course of antibiotics is taken.

✗ bacteria killed by antibiotics

🦠 bacteria resistant to antibiotics

Exam tip AQA

Apart from learning facts, you could be asked to examine data about infection rates or growth of bacterial cultures. Remember the rules for plotting graphs, and look for the trends in a graph or table.

Questions

1. Name two types of pathogen.

2. What does an antibiotic do?

3. Explain why scientists need to keep producing new types of antibiotics.

Revision objectives

- ✔ know how our bodies try to prevent infection
- ✔ describe how the immune system deals with pathogens
- ✔ know how immunisation protects against some diseases

Student book references

1.5 Immunity and immunisation

Specification key

- ✔ B1.1.2

Key words

immune system, antibodies, antigen, immunity, white blood cell, immunisation, vaccine

Exam tip   AQA

Make sure you are clear about the difference between the action of the phagocyte and the lymphocyte. Also be clear about the differences between natural and artificial immunity.

Questions

1 Name the two types of white blood cell.

2 What does immunity mean?

3 What is the difference between natural and artificial immunity?

Preventing infections

The body has a number of systems that prevent microorganisms getting in. These include:

skin – acts as a barrier

digestive system – stomach acid kills bacteria

blood clots – seal cuts

respiratory system – mucus traps bacteria.

The immune system

Once pathogens get inside the body the **immune system** comes to our rescue. It uses two types of **white blood cell**.

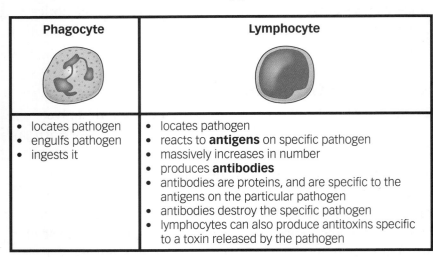

Phagocyte	Lymphocyte
• locates pathogen • engulfs pathogen • ingests it	• locates pathogen • reacts to **antigens** on specific pathogen • massively increases in number • produces **antibodies** • antibodies are proteins, and are specific to the antigens on the particular pathogen • antibodies destroy the specific pathogen • lymphocytes can also produce antitoxins specific to a toxin released by the pathogen

Immunisation

Immunity means the ability to resist an infection. This can be acquired in two ways.

Natural immunity

This happens when we are infected and some of the lymphocytes that produce the antibodies are retained to deal with future infections.

Artificial immunity

Here we are given a **vaccine**, which contains dead or inactivated pathogens. This triggers an immune response. The lymphocytes are retained, and will respond rapidly if a future infection by the pathogen occurs. For example, the MMR vaccine protects us against future infections of measles, mumps, and rubella.

If enough of the population becomes immune, the spread of infectious diseases is reduced. Unfortunately if a new strain develops by a mutation the vaccine will be ineffective.

Questions
Healthy living

Working to Grade E

1 What is the source of energy in a cell?

2 What is a balanced diet?

3 What is malnourishment?

4 What is the scientific term for being very overweight?

5 What are carbohydrates used for in the body?

6 What are fats used for in the body?

7 What are proteins used for in the body?

8 What is a pathogen?

9 What is a toxin?

10 Which are bigger, bacteria or viruses?

11 Name one harmful effect of bacteria in the body.

12 What is an antibiotic?

13 Name a painkiller.

14 Do painkillers kill pathogens?

15 Name a bacterium that is resistant to most antibiotics.

16 Name three barriers that prevent microorganisms entering the body.

17 Look at these drawings of blood cells.
   a Label the phagocyte and the lymphocyte.
   b What does a phagocyte do?

18 What is contained in a vaccine?

19 What does the MMR vaccine protect you against?

Working to Grade C

20 Suggest three things that might vary the metabolic rate.

21 Where does metabolism take place?

22 What is the relationship between the muscle-to-fat ratio in a person and their metabolic rate?

23 Name a health problem caused by being overweight.

24 Explain why it is important to wash our hands.

25 What do symptoms tell a doctor?

26 **a** Look at this diagram of the reproduction of viruses. Put the steps in the correct sequence.

The viral genes cause the host cell to make new viruses.

The genetic material from the virus is injected into the host cell.

The virus attaches to a specific host cell.

The host cell splits open, releasing the new viruses.

   b Use the diagram to describe the stages in the reproduction of a virus.

27 Explain why antibiotics have no effect on viruses.

28 What term is used for the chance changes that allow new strains of bacteria to develop resistance?

29 What does sterile mean?

30 Why is the apparatus used to grow microorganisms always sterilised before the set up of the experiment?

31 What is an inoculating loop used for?

32 Why are the lids of Petri dishes taped onto the base of the dish?

33 Why should cultures be incubated at 25 °C in a school laboratory?

34 Which organ system in your body deals with infections?

35 Describe how a lymphocyte destroys pathogens.

36 What is an antibody made of?

37 What is the difference between an antibody and an antitoxin?

38 What is immunisation?

Working to Grade A*

39 How might exercise change the muscle-to-fat ratio?

40 Explain how cholesterol might lead to a heart attack.

41 Explain how a bacterial infection makes you feel ill.

42 Explain how natural selection could lead to the development of antibiotic-resistant bacteria.

43 Explain why antibiotic resistance is of such concern to doctors.

44 What strategies do we use to reduce the development of resistant bacteria?

45 Explain why an antibody will only kill one type of bacteria.

46 Explain how a vaccination works.

Examination questions
Healthy living

1 The table below gives information about the nutrients (per 100 g of bar) in a number of chocolate snack bars and confectionery bars.

Type of bar	Energy (kJ)	Total fat (g)	Saturated fat (g)	Carbohydrates (g)	Sugars (g)	Proteins (g)
Milk chocolate bar	2200	30.0	18.6	56.8	56.6	7.5
Dark chocolate bar	2105	27.3	16.6	58.8	57.7	4.7
White chocolate bar	2315	33.3	20.6	59.7	59.7	4.5
Fruit and nut chocolate bar	2050	25.9	14.5	55.8	55.2	8.3
Raisin cereal bar	1800	15.7	8.6	65.1	42.3	5.4

a   Which bar will provide the most energy?

.... white chocolate bar ....

(1 mark)

b   Which **two** nutritional groups will provide the most energy?

...carbohydrate...and ...sugars............. .

(1 mark)

c   i   Which of the bars is the healthier bar to eat?

...............protein ✗...

(1 mark)

ii   Explain your answer.

.......protein is essential for cell repair, cell.............

.......replacement and cell growth......................................

....

(2 marks)
(Total marks: 5)

2   Look at the following graph, which shows the number of recorded cases of measles in the UK from 1940 to 2008. Measles was a common childhood disease caused by a virus.

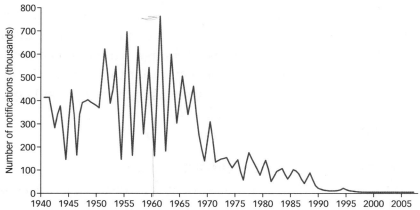

a   What is the trend in the graph for numbers of cases of measles?

During the years 1960 is reached its peak and the slowly retreated

(1 mark)

b   Two vaccines have been produced to combat measles. The first was a single vaccine. The second was the combined MMR vaccine.

   i   When do you think the first vaccine was introduced?

   ~~to test if the single vaccine~~ 1956

(1 mark)

   ii   Explain your reasoning.

   because from then one it decreased for 5 years ~~only~~ but then rose once again

(1 mark)

c   This treatment of measles is an example of active immunity. Explain how active immunity works.

The doctors inject you with a dead or inactive pathogen and allow your immune system to begin working, your body will restore the antibodies design that was released so if the same pathogen comes again you will body will deal with it faster.

(4 marks)

(Total marks: 7)

Revision objectives

- ✔ understand the structure and function of the nervous system
- ✔ recall the stages in a reflex arc
- ✔ know that many processes are coordinated by hormones
- ✔ know and understand why internal conditions are controlled

Coordination

Animals live in a changing environment. A change in the environment is called a **stimulus**. Animals need to respond to these changes by changing their behaviour. Selecting the appropriate behaviour for the stimulus is called **coordination**. There are two systems in the body that help coordinate:

- the nervous system
- the **hormone** system.

The nervous system

The nervous system is made up of nerve cells called **neurones**. It is divided into two parts:

- the **central nervous system** (CNS) – the brain and spinal cord
- the **peripheral nervous system** – the nerves taking messages to and from the CNS.

The nervous system acts by detecting the stimulus in **receptors** called sense organs. These send an electrical message called an impulse along nerves to the CNS. The CNS coordinates an appropriate response, then sends an impulse out to an **effector**. This brings about a response, by either contracting a muscle or releasing (**secreting**) a chemical substance from a gland.

Receptors

We have different receptors for the different stimuli. Some receptors, for example, the eye, have their own special cells. Others, like those in the skin, may just be the ends of the nerve cells. All have the basic parts of all animal cells: a cell membrane, nucleus, and cytoplasm.

Exam tip

Learn the following as the sequence or pathway of a nerve impulse:
Stimulus → receptor → sensory neurone → relay neurone → motor neurone → effector → response

Sense	Receptor	Stimulus
sight	eyes	light
hearing	ears	sound
balance	ears	changes in position
smell	nose	chemicals
taste	tongue	chemicals
touch	skin	pressure, pain, and temperature

Neurones

There are three types of neurone:

- Sensory neurone – takes impulses from receptors into the CNS.
- Relay neurone – takes the impulse from the sensory neurone to the correct motor neurone inside the CNS.
- Motor neurone – takes impulses from the CNS to the effector.

The junction between two neurones is called a **synapse**. The impulse passes across a synapse as a chemical message.

Reflexes

These are rapid, protective, automatic responses in the body. The pathway taken by the impulse is called a **reflex** arc. For example, pulling the hand away from a sharp pin is a reflex arc.

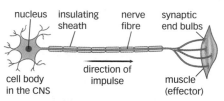

▲ Structure of a motor neurone. The nerve impulse is carried along the nerve fibre.

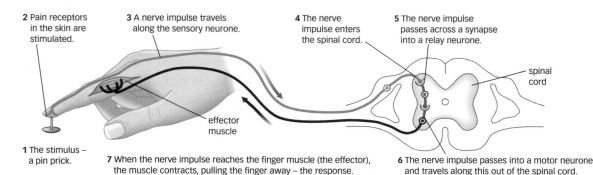

2 Pain receptors in the skin are stimulated.

3 A nerve impulse travels along the sensory neurone.

4 The nerve impulse enters the spinal cord.

5 The nerve impulse passes across a synapse into a relay neurone.

1 The stimulus – a pin prick.

7 When the nerve impulse reaches the finger muscle (the effector), the muscle contracts, pulling the finger away – the response.

6 The nerve impulse passes into a motor neurone and travels along this out of the spinal cord.

▲ A reflex arc. The impulse goes from receptor to CNS and then to effector to bring about the response. The relay neurone inside the spinal cord coordinates the response by connecting the sensory neurone to an appropriate motor neurone. The information travels from one neurone to another across a small gap called a synapse.

Hormonal control

The hormonal system plays a role in controlling the balance of our internal systems. It works by releasing hormones into the bloodstream. A hormone is a chemical messenger released in one **gland** and having its effect on a **target organ** elsewhere in the body. The effect is generally slower but longer lasting than nerve reflexes.

Internal conditions that are controlled

- Water content of the body – all cells in the body need to be bathed in water. Water enters the body in food and drink and is released during respiration. It leaves by the lungs when we breathe out, by the skin in sweat, and by the kidneys in urine.
- **Ion** content of the body – ions are needed to keep nerves and muscles healthy. Ions are taken in from the diet and lost in sweat and urine.
- Temperature – the body needs to maintain a stable temperature for its enzymes to function. The skin is particularly important in helping control temperature. For example, we sweat to cool us down.
- Blood sugar levels – sugar is needed as a source of energy for cells, but too much will lead to circulatory problems. Hormones regulate the level of blood sugar.

Key words

coordination, stimulus, central nervous system, peripheral nervous system, receptor, neurone, synapse, effector, reflex, hormone, secrete, gland, target organ, ion

Questions

1 What does coordination mean?

2 What is a receptor?

3 Explain the differences between nervous control systems and hormonal control systems.

Revision objectives

- understand the role of hormones in the menstrual cycle
- explain how oestrogen and progesterone are used in the contraceptive pill
- be aware that hormones are used to control fertility and in IVF treatment
- evaluate the benefits and difficulties of fertility treatments

Student book references

1.8 How hormones control the menstrual cycle

1.9 Using hormones to control fertility

Specification key

- B1.2.2

Hormonal control of the menstrual cycle

The **menstrual cycle** in females begins during puberty. The cycle is a sequence of events carefully controlled by **hormones**. It involves the exact timing of the release of an egg, and the preparation of the womb for pregnancy.

- The cycle begins with the release of a follicle stimulating hormone (**FSH**) from the pituitary gland.
- This stimulates the egg to mature in the ovary.
- FSH also stimulates the ovary to produce a hormone called **oestrogen**, which causes the wall of the womb to thicken.
- The high levels of oestrogen stimulate the release of a second pituitary hormone called luteinising hormone (**LH**), and stop production of FSH.
- On day 14 of the cycle the LH level peaks and causes a mature egg to be released; this is called **ovulation**.
- The ovary now produces a mix of oestrogen and **progesterone**.
- These hormones continue to thicken the wall of the womb, which can now receive a fertilised egg if the woman becomes pregnant.
- With high oestrogen and progesterone no FSH is produced, and no eggs mature.
- If the woman does not become pregnant, oestrogen and progesterone levels fall, and the wall of the womb is shed, together with some blood as the menstrual flow, or period.
- The cycle then starts again.

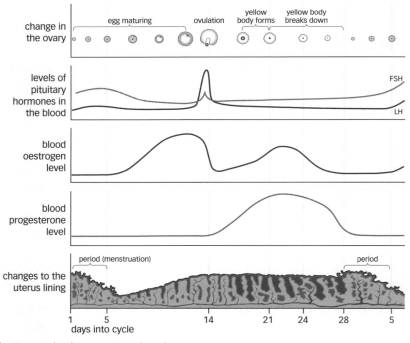

▲ Events in the menstrual cycle.

Controlling fertility

Our understanding of the hormones that control the menstrual cycle has allowed us to control **fertility** in humans.

Contraceptive treatments

Contraceptive drugs are ones that prevent pregnancy. These make use of the fact that the hormones oestrogen and progesterone will inhibit the production of FSH, and so stop eggs maturing.

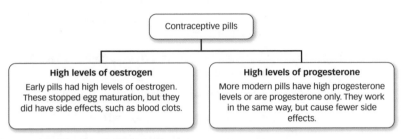

▲ Oestrogen and progesterone in the contraceptive pill.

Fertility treatments

Hormones can also be used to help women become pregnant. Some women have difficulty in maturing eggs, which may be because they have naturally low levels of FSH and LH. In such cases women can take a fertility drug, which contains both FSH and LH. This results in eggs being matured and released.

This treatment is useful in the process called **in vitro fertilisation (IVF)**.

- The woman is given a treatment of fertility drugs.
- Several eggs mature and are released; they are collected by the doctor.
- These eggs are mixed with sperm from the father.
- The fertilised eggs develop into small balls of cells, called embryos.
- One or two embryos are inserted back into the mother's womb (uterus).
- The embryos develop into a baby in the womb.

Controlling fertility in these ways has both benefits and problems.

- Some benefits include: families can control when they will start a family, and how many children they will have; IVF allows infertile couples to have children, and embryos can be screened for disorders.
- Some problems include: some people have ethical concerns about controlling fertility and disposing of unwanted embryos; IVF is expensive; there are side effects of the pill, and some women stay on the pill too long.

Exam tip

Remember that there are two pituitary hormones and two hormones made in the ovary involved in the menstrual cycle.

Questions

1. What two things does the menstrual cycle control?
2. Name two ways that doctors can use hormones to control the menstrual cycle.
3. **H** What is IVF?

17

Working to Grade E

1 Name three different stimuli the body can detect.

2 What stimulus is detected by the eye?

3 Which organ is sensitive to pressure?

4 Which two sense organs can detect chemicals?

5 What is an impulse?

6 Look at this drawing of a neurone.

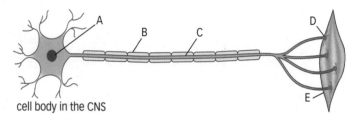

cell body in the CNS

    a Label parts A to E.
    b Identify the type of neurone.
    c Label the direction the impulse takes along the neurone.

7 What is a synapse?

8 What are the two parts of the nervous system?

9 Give an example of a reflex action.

10 Give an example of a voluntary action.

11 What secrete hormones?

12 Which organ regulates the amount of water in the body?

13 What is the normal temperature of a human body?

14 What is puberty?

15 Name the female sex hormones produced in the ovaries.

16 Name the hormone that stimulates the release of eggs from the ovary.

17 Where is FSH produced?

18 What does IVF stand for?

19 What is ovulation?

Working to Grade C

20 What is the function of a sensory neurone?

21 What is the function of a motor neurone?

22 How does the impulse pass across a synapse?

23 What is a reflex action?

24 Place the following structures in the correct order to describe the pathway of a reflex action:

motor neurone, receptor, stimulus, relay neurone, response, effector, sensory neurone

25 In which part of the nervous system do you find relay neurones?

26 How does a muscle respond to an impulse?

27 How are hormones transported?

28 Give three ways that water leaves the body.

29 Explain how the body cools down.

30 Name one process that generates heat in the body.

31 Give one way that ions pass out of the body.

32 Why is sugar needed by cells?

33 Where are hormones transported to in the body?

34 What is the function of FSH?

35 What two hormones are contained in contraceptive pills?

36 Which hormones are given to women who are having trouble conceiving?

37 Explain one medical complication caused by using hormones to control fertility.

38 What is the menstrual period?

Working to Grade A*

39 Explain why maintaining the body's temperature is important.

40 What stops the production of FSH during the menstrual cycle?

41 Which hormone combination in the contraceptive pill produces the fewest side effects?

42 Why are women who are about to undergo IVF given a course of hormone treatment?

43 List three benefits from using hormones to control fertility.

1 Sense organs detect stimuli. Connect the following sense organs to the stimuli they detect.

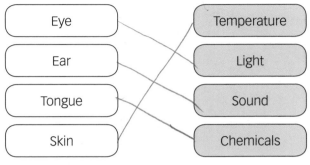

(4 marks)
(Total marks: 4)

2 What are the **two** component parts of the central nervous system (CNS)?

1 ....central nervous system..

2 ....peripheral nervous system..

(2 marks)
(Total marks: 2)

3 Hormones play a part in controlling female fertility.

a What is a hormone?

....a chemical messenge that travels through the
blood to a target organ....

(2 marks)

b In the female menstrual cycle:

i Which hormone triggers the maturation of the egg?

....~~Oestrogen~~ FSH....

(1 mark)

ii Which hormone triggers ovulation?

....Oestrogen....

(1 mark)

c Describe the role of hormones in the modern female **contraceptive pill**.

....Oestrogen inhibits the ~~growth~~ of production of
FSH and therefore stops an egg ~~entering~~ being released
into womb. Progesterone produces vaginal mucus
which ~~sto~~ acts as a barrie & stops sperm from entering
vagina....

(3 marks)
(Total marks: 7)

4 Below is a diagram that shows the structures involved in a spinal reflex.

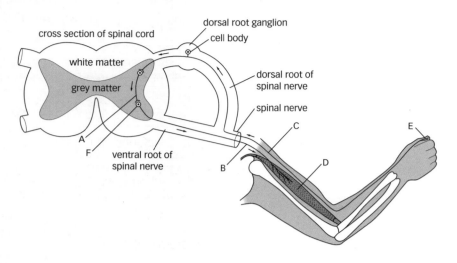

a Which of the structures labelled **A** to **F** is:

i the motor neurone? ...... B ....

ii the relay neurone? ...... A ....

iii the sensory neurone? ...... C ....

iv the sensory nerve ending? ...... E ....

stim → recep → fer → relay *(4 marks)*

b Describe how the labelled structures would be involved in the reflex after a drawing pin has entered the skin at **E**.

the sensory stimulus which is the drawing pin
would be ~~reced~~ detected by receptor & an impulse would be
sent through the sensory neurone to spinal cord
which would send the response through motor
neurone to the effector which is the muscle,
~~ordering~~ commanding it to contract

(6 marks)

c Give an example of where the brain can override the reflex action.

.... ~~it~~ before falling down ....

(1 mark)
(Total marks: 11)

5 A scientist carried out a test with a number of subjects to investigate human reaction times to a stimulus.
- Five subjects were selected.
- They were asked to play a computer game that had a button they could press every time they saw a red light on the screen.
- The computer showed the red light randomly a minimum of 10 times.
- The subjects' time to respond was recorded in milliseconds by the computer.
- For each game the computer calculated the average reaction time.
- The game was repeated five times for each subject.

Here are the results:

Subject	Average reaction time (ms)				
	Test 1	Test 2	Test 3	Test 4	Test 5
Nathan	120	110	109	101	98
Emily	157	122	118	101	108
James	143	133	136	126	109
Donna	128	98	103	97	90
Frankie	109	107	89	78	77

a What is the trend in the data?

After Test 1 the reaction times decrease for each subject

(1 mark)

b Which sense organ detected the stimulus?

eyes

(1 mark)

c How did the scientist attempt to make the experiment:

i accurate? _all we saw the were shown the red light for game was repeated 5 times_

ii reliable? _asked more than 1 or two people_

(2 marks)

d Emily thought her results might be wrong.

i Why does she think that? _because her test 5 was higher than test 4_

ii Would you agree with her or not? Explain your reasoning. _no, because it may just be an anoymu an anoynulus result. a fault_

(2 marks)

e Explain how you could adapt the experiment to show that people who regularly play computer games might react faster than people who do not play computer games.

....

....

....

....

....

....

....

(4 marks)
(Total marks: 10)

Plant hormones

Plants also need to respond to stimuli in the environment. Most plant responses are in the form of growth movements called **tropisms**. They respond to a number of stimuli.

Stimulus	Response of the shoot	Response of the root	Tropism
light	grows toward light	grows away from light	**phototropism**
gravity	grows away from gravity (up)	grows toward gravity (down)	**gravitropism (geotropism)**
moisture	grows toward moisture	grows away from moisture	hydrotropism

The growth movements in phototropism and gravitropism are brought about by a hormone called **auxin**. This is produced in the shoot and root tips, but acts behind the tip.

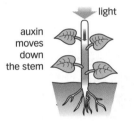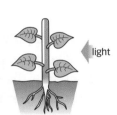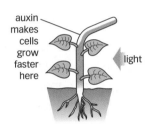

▲ Phototropism and auxin.

In the roots auxins slow plant cell elongation, and therefore the growth. In gravitropism the auxin sinks to the lower side of the plant, causing shoots to grow up away from gravity, and roots to grow down toward gravity.

Uses of plant hormones

Hormones are used commercially to affect plant growth. For example:

- weedkillers – here auxins cause the shoots of broad-leafed weeds to grow rapidly, but not the roots, so they will not be able to absorb enough water to survive.
- rooting hormones – here cuttings are dipped in rooting hormones containing auxins, which encourage the formation of new roots.

Exam tip AQA

Be able to predict the outcome of experiments involving plant growth in response to different directional stimuli. Be able to explain why these results occur in terms of auxin distribution.

Revision objectives

✔ know that plants respond to environmental stimuli such as light, gravity, and moisture
✔ understand the role of plant hormones in the control of growth
✔ understand the application of plant hormones in agriculture

Student book references

1.10 Controlling plant growth

Specification key

✔ B1.2.3

Key words

tropism, auxin, phototropism, gravitropism, geotropism

Questions

1 What is a tropism?

2 What do plants use to control tropisms?

3 Explain what happens to the distribution of auxin when light is shone on the plant from one side.

Revision objectives

- ✔ know what a drug is
- ✔ know that some drugs are useful
- ✔ know that medical drugs are tested before use
- ✔ know that there are different types of recreational drugs, some legal and some illegal

Student book references

1.11 Drugs and you

1.12 Testing new drugs

Specification key

- ✔ B1.3.1

Drugs

A **drug** is a chemical that affects our body chemistry. Scientists are continually developing new drugs. Drugs are used medicinally to treat illness but they are also abused by some people.

Use of medical drugs

Many drugs are developed to treat the causes and symptoms of illness. Beneficial drugs are painkillers, antibiotics, and **statins**. Statins are drugs that have been developed to reduce blood cholesterol level. These drugs reduce the incidence of cardiovascular disease, like heart attacks.

Drug testing

New drugs are constantly being developed. They need to be tested before they can be used by doctors to make sure that the drugs work, and that they are not dangerous.

During a **clinical trial** strict rules apply.
- Large numbers of patients give more reliable results.
- Start with low doses to check that the drug is safe.
- Increase the dose to find the best (optimum) dose.
- Some patients are given a **placebo** (this is a dummy pill without the active drug) to act as a control with which to compare the experimental group.
- **Double-blind trials** are used, where neither the patients nor the doctors know who has the placebo or the real drug, to avoid bias.

Thalidomide – a mistake to learn from

Thalidomide was a drug developed in the 1950s as a sleeping pill. It also helped treat morning sickness in pregnant women. Unfortunately the testing process was not thorough enough. They did not test it on enough types of animals or pregnant women. When they used the drug commercially with pregnant women, they discovered that it had side-effects. Many of the children born to mothers who took thalidomide had severe limb abnormalities. As a result the drug was banned. Scientists made their testing procedure far more rigorous.

Interestingly thalidomide is now used again, as a treatment for conditions like leprosy.

Drug testing

Step 1 – New drug is developed by scientists to treat a disease.

Step 2 – Laboratory trials, on cells, tissues, and eventually live animals. This will check for toxicity, and whether the drug works.

Step 3 – Clinical trials. Here the drugs are tested on human volunteers.

Step 4 – If successful the drug is marketed.

Use of recreational drugs

Some people use recreational drugs.
- Some drugs are legal and some are illegal.
- Some are more harmful than others.

Drug abuse is where people take drugs for no medical reason. Drugs act by changing the chemical processes in the body, particularly the brain. Some people develop an **addiction** to, or become dependent on, drugs. This means that they need the drugs to maintain a functioning lifestyle. If these people try to give up the drug, they suffer **withdrawal symptoms**, as the body's chemical reactions fail to function fully.

Legal recreational drugs

A commonly used legal drug is alcohol. In low doses this drug relaxes people and is not normally considered drug abuse. However, high doses of alcohol can impair judgement and lead to reckless behaviour, which might affect society as a whole. Long-term use of alcohol may lead to serious damage of the liver and other organs. Other legal recreational drugs include caffeine and nicotine.

Illegal recreational drugs

There are many types of illegal drug. For example:
- **Cannabis** – the smoke contains chemicals that give a feeling of well being. It is not significantly addictive, but tends to lead the user onto more powerful drugs. In some people it may lead to mental illness. However, it can be used to treat chronic painful illnesses such as multiple sclerosis.
- **Steroids** – these are performance-enhancing drugs, leading to muscle development. They give athletes an unfair advantage, and so are banned in sport. They can cause side-effects, such as interfering with the reproductive cycle and heart problems.

Other illegal drugs include cocaine and heroin. These are very powerful and harmful drugs. They are highly addictive.

Surprisingly, it is legal drugs, not illegal drugs, that have the greater impact on health. This is because more people use the legal drugs.

Key words

drug, statins, clinical trial, thalidomide, placebo, double blind trial, addiction, withdrawal symptoms, cannabis, steroids

Questions

1 Why do people find it difficult to stop taking some types of recreational drugs?

2 Why are large numbers of patients used in drug testing?

3 Why are laboratory trials of new drugs always carried out before clinical trials?

Working to Grade E

1   Name three things plants are sensitive to.

2   What is a stimulus?

3   Name a plant hormone.

4   What is a drug?

5   Give some examples of beneficial drugs.

6   Why do scientists carry out tests on new drugs?

7   How might drugs be tested in a laboratory?

8   What is a clinical trial?

9   What is a placebo?

10 Thalidomide is a drug.
    a   What was thalidomide used for?
    b   Thalidomide now has a new use. What is the new use of the drug?

11 What do statins do?

12 Define the following terms:
    a   drug abuse
    b   addiction
    c   withdrawal symptoms.

13 Some athletes use steroids as an illegal drug.

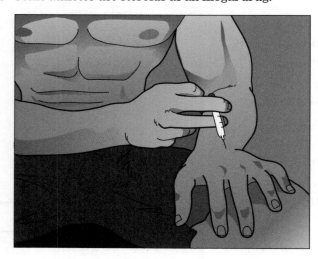

    a   Why might an athlete inject steroids?
    b   What medical problems could this cause for an athlete?

Working to Grade C

14 How do plants respond to stimuli?

15 Look at this drawing of bean seedling A.

    a   Draw the seedling as it would look after three days.
    b   Which part of the plant (the shoot or the root) responds positively to light?
    c   What is the name of this response?

16 Look at this drawing of bean seedling B.

    a   Draw the seedling as it will look after three days in an unlit box.
    b   What stimulus is the bean seedling responding to?
    c   What is the name of this type of response?

17 Name one other commercial use of plant hormones (other than weedkillers).

18 a   What is the purpose of a double-blind trial?
    b   How is a double-blind trial carried out?

19 What is a side-effect?

20 Which diseases do statins help to prevent?

21 A doctor carried out an experiment using patients on a motorbike simulator. They had to brake when the screen showed a person crossing the road. The patients then had to drink alcohol, and repeat the test. The results are shown below.

Alcohol concentration (mg/l)	Average braking time (s)
0	0.572
0.15	0.585
0.25	0.610

    a   Describe the pattern in the data.
    b   What can you conclude about the effect of alcohol on people's reaction times?

c How would a study like this improve road safety?

d How could you take the experiment further to establish the point at which alcohol might start to have an effect?

e How could you design an experiment using a placebo to support the view that alcohol is having an effect?

Working to Grade A*

22 Explain why steroids are banned in sport.

23 What is the effect of auxins on plant cells in the shoot?

24 Explain how auxins function as a weedkiller.

25 Explain how auxins cause a plant to grow toward the light.

26 Look at the diagram below of cress seedlings growing on a piece of equipment called a clinostat. The equipment rotates one full circle in an hour.

a Predict what the seedlings will look like after three days.

b Explain why this has happened.

c What would happen if the clinostat stopped working after a few hours?

27 Astronauts carried out experiments on plant growth in space. They grew bean seedlings in dark boxes for several days. Below is a drawing of the results of such an experiment.

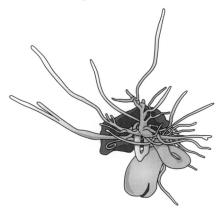

a Can you describe any clear patterns of directional growth in the seedlings?

b Explain why such results were achieved.

28 Suggest how you might modify the experiment to achieve some kind of directional growth.

29 **a** What problems did the drug thalidomide cause?

b What was wrong during the development process of thalidomide, which resulted in such problems?

c What precaution would a doctor take before prescribing thalidomide now?

30 Discuss the benefits of and problems with the use of cannabis.

31 A newspaper has suggested that illegal drugs such as cannabis and heroin are the most significant cause of drug-related health problems in the UK. Do you agree with this claim? What evidence can you use to support your answer?

32 Explain why someone might become dependent on a drug.

1 People use drugs for many different reasons.

a What is a drug?

a chemical that affects our bodies chemistry

(1 mark)

b Give an example of a medical drug.

antibiotic

(1 mark)

c Give an example of a recreational drug.

cannabis

(1 mark)

d Scientists now think that legal recreational drugs cause more problems than illegal recreational drugs. Explain why this is the case.

Because they are legal they are sold in stores which makes them more accessable

(1 mark)

e What is drug addiction?

when somebody's body can't function properly without taking a drug

(2 marks)

(Total marks: 6)

2 Plants show growth responses to environmental stimuli.

a Name the response that causes roots to grow down with gravity.

gravitropism

(1 mark)

b Explain why it is an advantage to the plants for the shoots to show phototropism.

it is able to grow & take in more sunlight for photosynthesis

(2 marks)

c Explain how phototropism is caused in the shoot.

Auxins are at the tip of shoots & roots Auxins accumulate at the shadest side of a shoot & elongate that part so the plant grows towards light

(2 marks)

(Total marks: 5)

Presenting and using data

Within this module there are some very good examples of the presentation and interpretation of data. You may be asked in a question to demonstrate these skills yourself.

Look at the following example; this will show you how scientists present and interpret data. There are opportunities for you to practise some of the skills, and to be guided through some of the more difficult elements that might appear in a question in the exam.

The flu pandemic of 2009

During the autumn of 2009, the world faced a new form of the flu virus (H1N1, also known as 'swine flu') that had originated in Mexico. Scientists all over the world were tracking the spread and effects of this disease. The aim was to be able 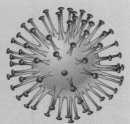 to prepare countries for the arrival of the disease, and give them some indication of how it would spread.

Doctors in the UK collected data on the number of cases. This was done in every health region of the UK. The following data is for the city of Sheffield.

Week beginning	Number of swine flu cases per 100 000 of the population
14 September	8.1
21 September	28.8
28 September	35.5
5 October	50.3
12 October	50.9
19 October	46.9
26 October	52.9
2 November	70.8
9 November	33.3
16 November	22.5

1   Plot the data on a graph. Join the points.

Skill – Presenting the data

This kind of question shows how scientists try to illustrate complex data as an image (the graph). Graphs give a picture of the data and are useful for identifying patterns.

The plotting of a graph is a common exam question, but it is surprising how many students drop marks on these sorts of questions.

Here are some quick, easy tips for plotting graphs: remember the word 'SLAP':

- **S** = Scales – when spreading the scales on the axes, the values on the axes must go up in equal amounts for each square size on the graph.
- **L** = Labels – remember to label each axis with the variable and the unit.
- **A** = Axes – make sure these are plotted the correct way round: the independent variable (the one you set or change, such as 'week beginning') along the 'X' axis, and the dependent variable (the one you measure, such as 'number of cases of flu') up the 'Y' axis.
- **P** = Plotting the points – your points should be plotted with a small neat 'x' in the exact position.

Finally, remember to join the points.

Skill – Using data to draw conclusions

Once scientists have plotted the graph they can start to make sense of complex data. There are a few common types of question that scientists ask, and you could be asked these same questions in an exam.

2   What is the trend shown by the data for flu cases in Sheffield?

This type of question requires an overview of the pattern shown by the graph. Describe the overall rise/fall in the line of the graph.

Common mistakes:
- It does not require a point-by-point description.
- Do not describe the first half only and then ignore the second half, where the direction or trend might change.

3   Are there any anomalous results?

This question asks you to pick up any results that do not fit with the trend you have described. Look for any point that stands out as unusual or that distorts the shape of the curve on the graph.

4   When does the flu epidemic appear to be under control?

This is asking you for the point on the graph where the line changes direction.

5   What might have been the result for the number of cases on 23rd November?

This question asks you to continue the shape of the graph a little further, and to take a reading at that point. There is usually a little flexibility for the examiner to allow a small range in the answers, as there is no point on the graph.

AQA Upgrade

Answering a question with data response

Tuberculosis (TB) is a disease caused by a bacterium.

1 How does the use of antibiotics help a patient with TB? *(1 mark)*

2 This graph shows the number of cases of TB in England and Wales since 1930. What trend is shown by the data? *(2 marks)*

3 Has the introduction of a vaccine eliminated the disease? Explain your answer using evidence. *(2 marks)*

4 By the 1970s many TB wards were closed. Suggest the reason for this. *(1 mark)*

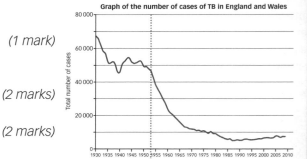

Graph of the number of cases of TB in England and Wales

G–E

1 Antibiotics fight infections.

2 In 1952 the number of cases was about 50 000.

3 No, because I can see on the graph.

4 There was a lack of money to keep them open.

Examiner: The candidate gives a vague answer, which does not describe what an antibiotic does. It is not worth a mark.

The candidate has not given a trend, but picked on one point of data. They have put effort into looking at the graph, but will not get any marks because the trend has not been given.

This candidate has got one mark for recognising that there are still cases. They have tried to use the graph, but not explained what it tells them and how. You need to quote figures or trends directly from the graph to get these marks.

No marks given because the candidate has not used the graph to suggest an explanation.

D–C

1 Antibiotics kill bacteria and viruses.

2 There has been a gradual fall in the number of cases.

3 No. People still get TB.

4 The cases have fallen.

Examiner: The answer is slightly muddled or confused. It shows some understanding of the action of antibiotics, but the use of the word virus confuses the answer, as viruses are not killed by antibiotics. No mark can be awarded, as the use of the word viruses is incorrect.

The decline has been recognised, but the significance of the introduction of the vaccine has not. This will only gain one of the two marks.

The candidate has got a mark for recognising that TB is not eliminated, but will not get the second mark because they have not referred to the data in the graph.

Whilst the candidate has recognised the decline in numbers, they have not read from the graph that the number of cases was now low. Therefore no mark can be awarded.

B–A*

1 Antibiotics kill bacteria.

2 There has been a decline in the number of cases, but there was a significant drop after 1953 when the vaccine was introduced.

3 No, because the graph still shows about 5–10 000 cases of TB even today.

4 The number of cases were so low, there was no need for them.

Examiner: Clear and accurate.

An excellent answer, not only picking out the overall trend, but also the significant drop after the vaccine.

The student has given a clear answer to both parts of the question.

Again clear answer. The candidate has recognised that the important point is that the case number being low meant that there was no need for them.

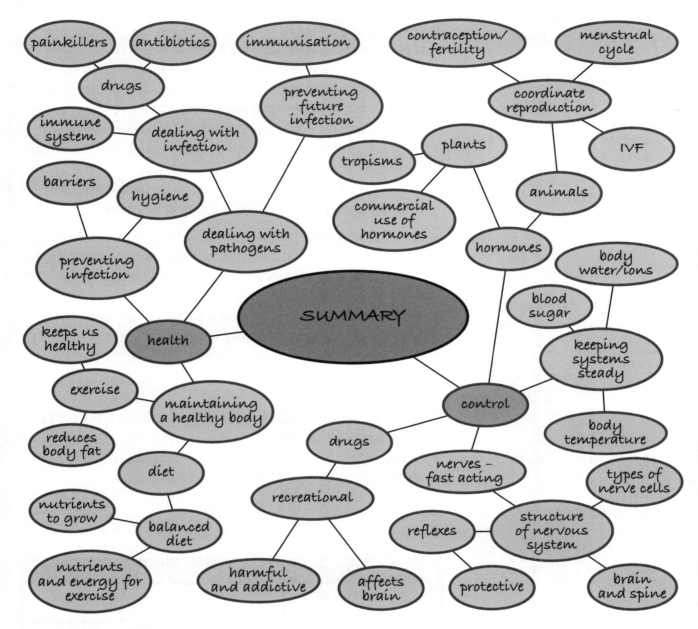

Revision checklist

- To stay healthy humans need a balanced diet. This provides enough of each nutrient and the energy for the body's needs.
- The amount of energy needed depends on your level of activity and metabolic rate.
- Pathogens are organisms, mainly bacteria and viruses, that can infect our body, multiply, and make us ill.
- Our body has a number of barriers that prevent pathogens getting into the blood.
- Our white blood cells are able to fight infections by either ingesting pathogens or making antibodies to kill them. Antitoxins are made to remove toxins.
- Vaccines can generate immunity to pathogens in our bodies.
- Drugs like antibiotics kill bacteria; painkillers can remove painful symptoms. Bacteria can develop resistance to antibiotics.
- Our bodies have a nervous system. The brain and spinal cord coordinate the correct response to any stimulus.

- Reflexes are rapid, automatic, and protective responses.
- The body is able to control its temperature, blood sugar levels, water content, and ion content at a steady level.
- Hormones are chemical messengers that coordinate the actions of cells.
- Hormones are used to control a woman's menstrual cycle. Knowledge of this can be used to control fertility and contraception.
- Plants use hormones to respond to changes in their environment. These hormones are used in agriculture to control growth.
- Drugs change the way the body or brain works. They may be medicinal or recreational. Some drugs are harmful or addictive.
- New drugs must be tested thoroughly before they can be used on patients.

Revision objectives

- ✔ know that adaptations of organisms help them to survive
- ✔ identify and explain key adaptations of plants and animals to cold or dry environments
- ✔ understand that the adaptations of an organism determine where they can live
- ✔ know that organisms live in extreme environments and show adaptations to those environments

Being adapted

Successful plants and animals are well suited to their environments. They show **adaptations**. An adaptation is any feature that aids survival and reproduction. Adaptations help the organism to compete with others for limited resources:

- Adaptations may help gain materials and resources from the environment. For example, large leaves can absorb more sunlight.
- They may help gain materials from other organisms. For example, sharp teeth will help kill prey animals.

Some adaptations may be very specific to a particular feature of the environment or the organism's way of life. For example:

- Thorns on some plants, such as the acacia tree, reduce grazing by animals.
- Poisons in some plants or animals, such as the poison dart frog, reduce grazing or predation.
- Warning colours, such as the yellow and black stripes of a bee, deter predators.

Animals are adapted to their environment

Animals are successfully adapted to survive in environments from the arctic to the desert. For every adaptation you should be able to give a reason why it aids survival.

Adaptations to arctic conditions

Here are the key adaptations to remember:

Adaptation	How this aids survival
changes to **surface area**, for example, small ears	This reduced surface area reduces heat loss.
thickness of insulating coat	The thick fur coat insulates the body against the cold.
amount of body fat	The bear has a thick layer of fat, which insulates against heat loss, and can be used in respiration to generate heat.
camouflage	The white fur means that the animal blends in with the environment.

Exam tip

The exam question could be about any plant or animal. Practise looking at any plant or animal, identifying an adaptation, and suggesting how it might aid that plant or animal to survive.

Adaptations to desert conditions

Here are the key adaptations to remember:

Adaptation	How this aids survival
changes to surface area, for example, long legs	This lifts the body high above the hot sand.
thickness of insulating coat	The thin in fur coat traps less insulating air.
amount of body fat	The thin layer of body fat reduces heat retention. However, there is a store of fat in the hump that can be used to release energy and water.
camouflage	Sandy coloured fur blends in with the background.

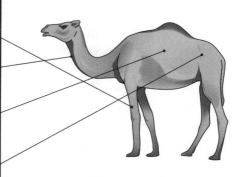

Plants are adapted to their environment

Plants are also successful in a wide range of environments. One of the most challenging environments must surely be the hot, dry environment of the desert.

Here are the key adaptations to remember:

Adaptation	How this aids survival
changes to the surface area of the leaves	Leaves are **spines**, which reduces surface area. This in turn reduces water loss.
water-storage tissues	Stems are swollen to store water.
extensive root systems	The roots of a cactus are often shallow but they cover a large area, which allows a greater absorption of water when it does rain.

Adaptations to extreme environments

Many organisms are able to withstand extreme conditions. They are called **extremophiles**. Again, they show adaptations to help them survive under such conditions.

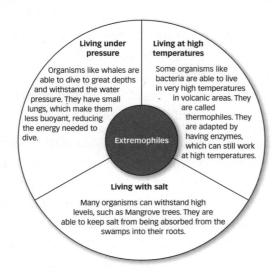

Living under pressure

Organisms like whales are able to dive to great depths and withstand the water pressure. They have small lungs, which make them less buoyant, reducing the energy needed to dive.

Living at high temperatures

Some organisms like bacteria are able to live in very high temperatures in volcanic areas. They are called thermophiles. They are adapted by having enzymes, which can still work at high temperatures.

Extremophiles

Living with salt

Many organisms can withstand high levels, such as Mangrove trees. They are able to keep salt from being absorbed from the swamps into their roots.

Questions

1 What is an adaptation?

2 Why are there so few successful plants and animals living in hot, dry deserts?

3 **H** What is the major problem for animals living in an arctic environment? Suggest one way the arctic fox is adapted to survive the conditions.

Revision objectives

✓ know that organisms compete with each other for resources and that this can affect their distribution
✓ know that the distribution of organisms is affected by changes in the environment
✓ explain how organisms are affected by pollution
✓ identify how organisms and apparatus can be used to indicate levels of pollution

Student book references

1.16 Competition for resources
1.17 Changes in distribution
1.18 Indicating pollution

Specification key

✓ B1.4.2

Distribution of an organism

A **population** is the number of individuals of a species in a named area. The **distribution** of the population is the range and extent of the area in which it lives. The distribution can be affected by:

- **competition** with other organisms
- effective adaptation to the environment
- changes in the environment
- pollution.

Competition

Organisms need **resources** to survive. In any environment the resources are limited. All organisms compete with each other for those resources.

Plants compete for: *Animals compete for:*

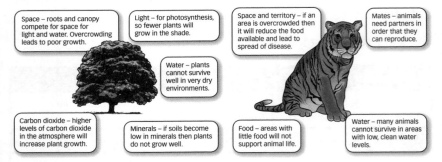

Space – roots and canopy compete for space for light and water. Overcrowding leads to poor growth.

Light – for photosynthesis, so fewer plants will grow in the shade.

Water – plants cannot survive well in very dry environments.

Carbon dioxide – higher levels of carbon dioxide in the atmosphere will increase plant growth.

Minerals – if soils become low in minerals then plants do not grow well.

Space and territory – if an area is overcrowded then it will reduce the food available and lead to spread of disease.

Mates – animals need partners in order that they can reproduce.

Food – areas with little food will not support animal life.

Water – many animals cannot survive in areas with low, clean water levels.

Adaptation

A well-adapted organism survives well in its environment. It usually means that it can obtain the resources it needs to survive.

Changes in the environment

Being well adapted to an environment causes problems if the environment changes. This usually means that an organism has to change its distribution, finding a new area to which it is better suited. Changes might be caused by:

- the arrival of competitors
- changes in the physical conditions in the environment, such as temperature or rainfall.

A good example of this is seen in various bird migration patterns. The ringed plover lives in Scandinavia during the summer, and used to migrate to Britain to overwinter in a milder climate. However, the climate has become warmer in mainland Europe during the winter. This has resulted in a change to the migration pattern, and many plovers now migrate to Europe rather than Britain during the winter.

▲ Ringed plover.

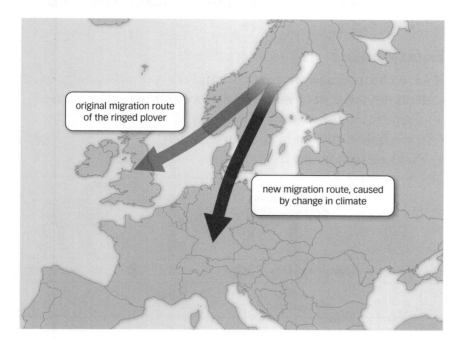

◀ Climate change has resulted in
lower numbers of ringed plover in
the UK.

Pollution

Pollution is the release of harmful substances into the
environment by humans. The substances are called **pollutants**.
Pollutants affect the survival and distribution of organisms.

- Some species can survive well in high levels of pollution.
  Their presence indicates to biologists that the area is polluted.
- Other species only survive in clean areas. Their presence
  indicates that the area has no pollution.

Such species are known as **indicator species**.

Indicators of air pollution

Lichens are very good indicators of pollution levels in the air.
Some species are particularly sensitive to sulfur dioxide.
Sulfur dioxide is released as a pollutant from burning fuels.
These lichens don't grow in industrial areas.

Indicators of water pollution

Polluted water contains high levels of microorganisms, which
tend to massively reduce the level of oxygen in the water.
Invertebrates like the rat-tailed maggot are able to survive well in
polluted waters that have low oxygen levels. This is because they
have a straw-like tail that can obtain oxygen from the air. The
presence of these maggots indicates that the water is polluted.

Measuring pollution levels

Apart from using indicator species, scientists can also use a
range of sensors that can measure physical or chemical levels
in the environment. These include maximum–minimum
thermometers, oxygen meters, and rainfall gauges.

Exam tip AQA

Exam questions are often based
on information about a named
species. Look out for information
about the effects of competition,
changes in the environment, and
pollution on the species, and be
ready to discuss how it might
affect the species' distribution.

Questions

1   What is competition?

2   What might affect the
    distribution of an animal?

3   **H** How might pollution
    affect the distribution of
    organisms?

Revision objectives

- ✔ understand how food chains and pyramids of biomass show the feeding relationships between organisms
- ✔ know that energy and biomass are lost at every link in the food chain
- ✔ appreciate that farming methods try to reduce energy loss

Student book references

1.19 Pyramids of biomass

1.20 Energy flow in food chains

Specification key

✔ B1.5.1

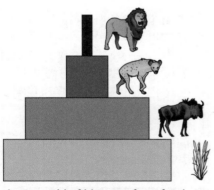

▲ Pyramid of biomass for a food chain on the African savannah.

Exam tip AQA

Learn the rules of plotting pyramids of biomass. You may be asked to plot or interpret these pyramids.

Food chains

A **food chain** shows what eats what.

- It shows the flow of **energy** and food (**biomass**) from one organism to the next.
- Each link in the chain is given a name.
- The links are joined by an arrow.

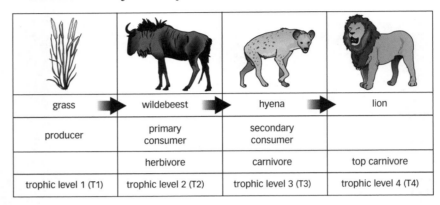

grass	wildebeest	hyena	lion
producer	primary consumer	secondary consumer	
	herbivore	carnivore	top carnivore
trophic level 1 (T1)	trophic level 2 (T2)	trophic level 3 (T3)	trophic level 4 (T4)

Biomass

Biomass is the mass of living material. The biomass of each link in the food chain can be calculated by multiplying the number of individuals in that link by the dry mass of one individual. Biologists then plot this as a **pyramid of biomass**.

Pyramids of biomass

Rules for plotting a pyramid of biomass:

- The producer is always at the base.
- Each bar represents the biomass of each trophic level.
- Each bar must be drawn to the same scale.

When plotted a typical pyramid shape is usually produced. This tells biologists that at each stage in the food chain there is less biomass than at the stage before. Biomass is lost because not all of the organism is eaten, and because some of the material eaten is lost in the waste droppings.

Energy flow through the food chain

Food chains don't only show the movement of biomass from one organism to another, but also the energy. As with biomass, some energy is lost at each link in the food chain.

Energy efficiency in farming

Many modern farmers produce meat for the human food chain. Since energy (and biomass) is lost at every link in the food chain, our scientific understanding of food chains can make the process more energy efficient.

Sun

Light radiated from the Sun is the source of energy for the food chain.

Green plants and algae absorb a small amount of the light energy (about 1%) during **photosynthesis**.

Most of the light energy (about 99%) is reflected.

Some of the absorbed energy is lost as heat to the surroundings.

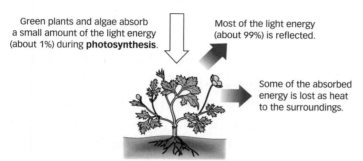

The remaining absorbed light energy is converted to chemical energy and stored in plant compounds.

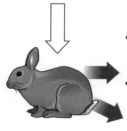

• Some plant compounds are used in respiration to supply the energy needs of the animal, including movement.
• During this transfer of energy, much is lost as heat to the surroundings.

Some energy lost in feces.

Some energy is built into animal compounds.

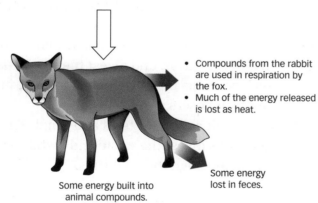

• Compounds from the rabbit are used in respiration by the fox.
• Much of the energy released is lost as heat.

Some energy lost in feces.

Some energy built into animal compounds.

▲ Energy flows through a food chain. Some energy is transferred out at each stage.

- Farming food chains are short – fewer links mean smaller losses.
- Animals kept warm – less heat energy is transferred to the surroundings.
- Animals enclosed in small pens – reduces energy loss from movement.
- Pests reduced – these compete with the organism for available energy sources.

Key words

food chain, biomass, pyramid of biomass, energy, photosynthesis, respiration

Exam tip AQA

This energy-loss diagram can look complicated. Focus on the loss of energy at each step, and notice that the pattern for all the animals is very similar.

Questions

1 What is the initial source of energy in all food chains?

2 What is biomass?

3 H How does intensive farming reduce energy loss through the food chain?

Revision objectives

✓ know that nature recycles by the decay of dead material
✓ understand the important part microorganisms play in the process of decay
✓ explain that elements are cycled between the living and non-living world
✓ understand the steps in the carbon cycle

Student book references

1.21 Recycling in nature
1.22 The carbon cycle

Specification key

✓ B1.6.1, ✓ B1.6.2

Recycling

Living things are made of energy and materials. We have seen that energy moves into and through the living world, in the

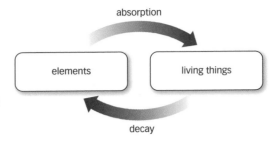

food chain. The materials are made of elements that are in a constant cycle between the living and non-living world.

When an organism dies, its body **decays** and returns the elements to the non-living world, like the soil and the air. They then become available for other organisms, such as plants, to absorb and use. This is natural **recycling**.

A stable community, like a woodland, does not require any input of materials, just a source of energy from the Sun to enable **photosynthesis**. This means that there must be a constant recycling of the elements. There must be a balance between the absorption of the elements and the return of the elements by decay.

Decay

Decay is where the waste from an organism, or the dead body of an organism, is broken down. There are two principal groups of organisms involved in decay:

- Detritivores – including earthworms, which eat bits of dead body such as dead leaves, and digest them, releasing their waste, which contains the elements and can be broken down further by decomposers.
- Decomposers – including bacteria and fungi. These **microorganisms** digest the waste and bodies by releasing enzymes to break down the materials.

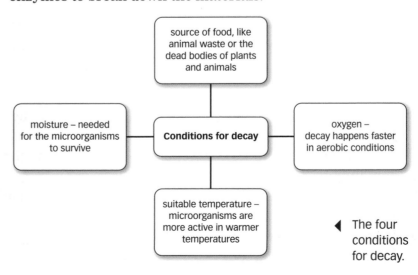

◀ The four conditions for decay.

The carbon cycle

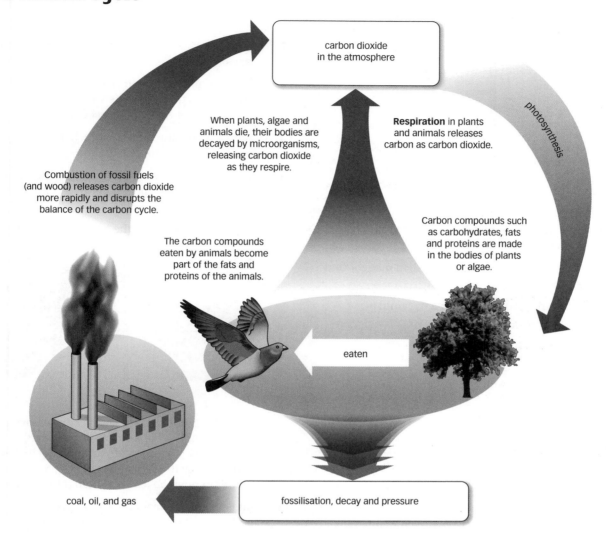

carbon dioxide in the atmosphere

When plants, algae and animals die, their bodies are decayed by microorganisms, releasing carbon dioxide as they respire.

Respiration in plants and animals releases carbon as carbon dioxide.

photosynthesis

Combustion of fossil fuels (and wood) releases carbon dioxide more rapidly and disrupts the balance of the carbon cycle.

Carbon compounds such as carbohydrates, fats and proteins are made in the bodies of plants or algae.

The carbon compounds eaten by animals become part of the fats and proteins of the animals.

eaten

coal, oil, and gas

fossilisation, decay and pressure

▲ The major steps in the carbon cycle.

During the **carbon cycle**, the element carbon is constantly being cycled between the living and non-living world. While the elements cycle, energy goes through the cycle. When carbon dioxide is built into carbohydrates and fats in plants, they take in and store the Sun's energy in these compounds. As the compounds move through the cycle, the energy either passes from one step to the next, or is released back into the atmosphere.

Exam tip

The examiner often asks you to identify where the key biological processes, like photosynthesis, occur in the carbon cycle. Learn the steps in the cycle, focusing on the points at which the biological processes occur.

Questions

1 What is decay?

2 What is the carbon cycle?

3 Explain how the action of detritivores speeds up the process of decay.

Key words

microorganism, decay, recycling, carbon cycle, photosynthesis, respiration, combustion

Working to Grade E

1 Give one way a fish is adapted to living in water.

2 Explain how a cactus is well adapted to living in dry conditions.

3 Name two extreme conditions in which microorganisms are found.

4 Sharks and dolphins have a streamlined body shape. What does this phrase mean?

5 List the five resources that plants compete for.

6 What is pollution?

7 Name one air pollutant and state where this pollutant comes from.

8 Describe one adaptation of a rat-tailed maggot that allows it to live in polluted water.

9 Where does the producer go in a pyramid of biomass?

10 Apart from energy, what is lost at every link in the food chain?

11 What is a trophic level?

12 What is a detritivore?

13 Name two decomposers.

14 Name the biological process by which living things release carbon dioxide back into the atmosphere.

15 Name three compounds that contain carbon in the body of a plant.

16 How do carbon compounds pass from plants into animals?

Working to Grade C

17 Why is a shark's streamlined body an adaptation to its way of life?

18 Whales are adapted to survive at high pressures.
   a What does this allow the whale to do?
   b Explain why they might need to do this.

19 Explain why the air spaces in the bones of birds are a useful adaptation for flight.

20 What is the relationship between the thickness of a mammal's fur and the habitat that it might live in?

21 Look at the drawings of the two plants below.

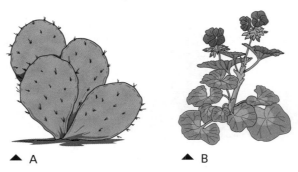

▲ A   ▲ B

   a Which plant lives in a hot and dry environment and which lives in a wet environment?
   b Identify two features of A that make it well adapted for its environment.
   c Explain how the features help it to survive.

22 What is the name given to all organisms that scientists use to work out whether an area is polluted?

23 Look at this graph containing data from a river with some pollution.

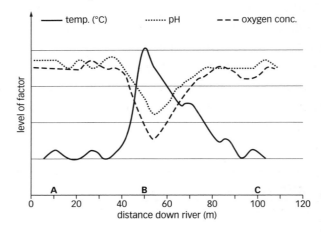

   a At what point along the river is the greatest level of pollution?
   b Explain the oxygen readings at this point.

24 What is light energy converted to in plants?

25 During what process does this conversion occur?

26 Explain why it is more energy efficient for a farmer to keep a herd of cows in a small barn, rather than roaming in large fields.

27 What data would you need to collect to construct a pyramid of biomass?

28 Look at the food chain below.

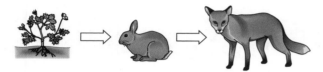

a Sketch a pyramid of biomass for this food chain.
b Explain the size of the bar for the fox in the pyramid.
c Energy is lost at each link in the food chain. List three ways in which the energy is used up.
d Explain how this loss of energy limits the length of the food chain.

29 An ideal environment, like a natural woodland, is called a stable environment. What must happen to the amount of carbon dioxide released and the amount absorbed for the environment to be stable?

30 Describe the process by which fossil fuels are formed.

31 Explain how the process of combustion upsets stable environments.

32 A biologist carried out an experiment to measure the rate of decay in bread. Below are the results.

Conditions	Time for mould to appear on bread (days)
Dry/cold	20
Dry/warm	10
Wet/cold	8
Wet/warm	3

a Which conditions were best for decay?
b Explain the result for the dry and warm condition.
c How could the results be made more reliable?

Working to Grade A*

33 Look again at the plants in question 21. Predict the outcome if plant B was planted in the environment of plant A, and explain your reasoning.

34 Biologists are able to grow human gut bacteria and thermophile bacteria in the laboratory.
a On which organisms would they choose to carry out genetic engineering experiments at 50°C?
b Explain your choice.

35 Look again at the graph of river pollution in question 23. In zone A on the river there are many mayfly larvae but few rat-tailed maggots. Suggest a reason for this.

36 Biologists believe that red and grey squirrels are in direct competition with each other, resulting in a change in the distribution and population size of the red squirrel. Suggest three pieces of evidence biologists would need to collect to support this theory.

37 In a recent survey, a team of biologists measured the levels of pollution in an inner city in England. They recorded the number and types of lichens, and took measurements of several air pollutants. They concluded that the pollution had been present for many years.
a What pieces of evidence allowed the biologists to reach this conclusion?
b Explain why you have selected this evidence.

38 Look again at the table measuring the rate of decay in bread in question 32.
a Based on these results what advice could you give about the best conditions to store bread?
b How could you modify this experiment to establish the ideal temperature for bread to decay?
c In this experiment, identify two variables you would have to keep the same.

39 Some biologists believe that planting forests of trees will offset our carbon emissions. What does this mean?

40 Explain how carbon found in molecules in a dead animal might become available for a plant and end up built into the body of a plant.

1 Below is a drawing of a ring-tailed lemur. They live in the trees on the island of Madagascar.

a Use the information in the drawing to suggest **one** way in which it is adapted to life in trees.

....

....

(1 mark)

b There are several different species of lemur living in the forests of Madagascar. Below is a table of data collected by biologists studying the three lemur species.

	Ring-tailed lemur	Mouse lemur	Common brown lemur
Time of main activity	daytime	nocturnal	daytime
Type of vision	colour	black and white	colour
Diet	mainly leaves and herbs	mainly fruit and insects	mainly bark

Use the information in the table to suggest **two** reasons why the ring-tailed lemur and the mouse lemur do not compete with each other even though they live in the same environment.

....

....

....

(2 marks)

c All three species of lemur are classified into the same major group. Explain how biologists are able to group them together in this way.

....

....

....

(2 marks)

d Biologists believe that these three species are closely related. How can different species be related?

....

....

(1 mark)

(Total marks: 6)

2 Below is a diagram of the carbon cycle.

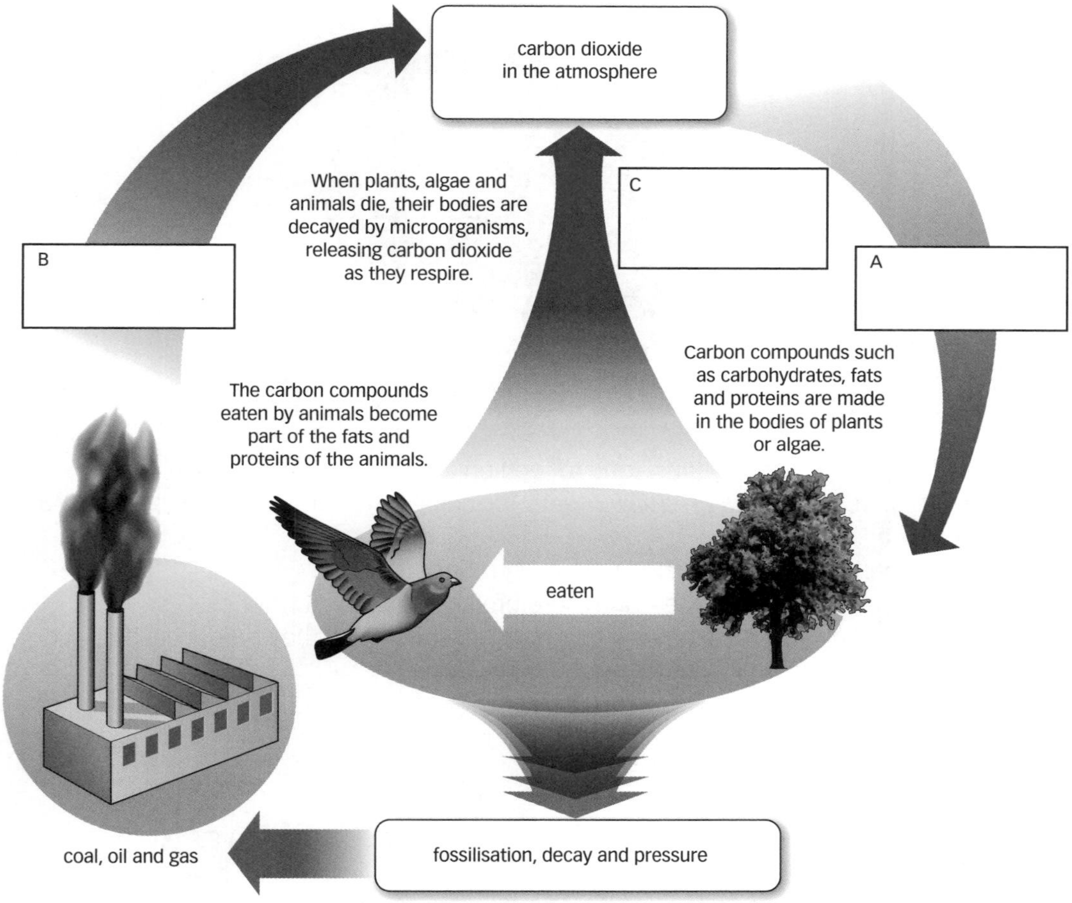

carbon dioxide in the atmosphere

When plants, algae and animals die, their bodies are decayed by microorganisms, releasing carbon dioxide as they respire.

C

B

A

Carbon compounds such as carbohydrates, fats and proteins are made in the bodies of plants or algae.

The carbon compounds eaten by animals become part of the fats and proteins of the animals.

eaten

coal, oil and gas

fossilisation, decay and pressure

a Name processes A, B, and C by writing them in the correct boxes in the diagram. *(3 marks)*

b In the carbon cycle decay is an important process that returns carbon dioxide to the air.

Which groups of organisms are involved in the decay process?

....

....

(2 marks)

c There are a number of factors that affect the rate of decay. State **two** factors that affect the rate and explain how they affect the rate.

....

....

....

....

....

....

(4 marks)

(Total marks: 9)

3 Ladybirds are common in Britain. There are 46 species in the UK. In 2004 the Harlequin ladybird arrived in Britain from northwest Europe.

- Harlequin ladybirds, like most ladybirds, tend to feed on aphids.
- The Harlequin ladybirds also feed on small insects, including other ladybirds, the eggs and larvae of butterflies, pollen, and nectar.
- They have longer periods of reproduction than most other species.
- They can fly rapidly over long distances, and therefore arrive in new territories.
- Sightings of this invading species are being logged by scientists and they produce distribution maps.

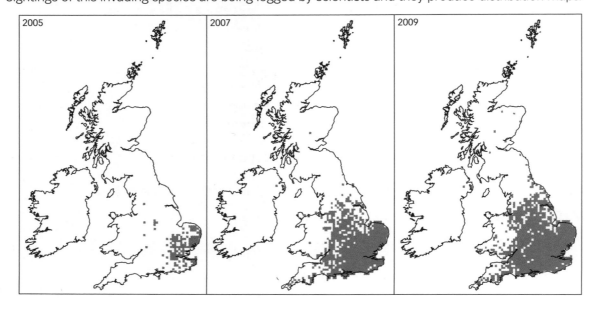

a What pattern can you see in the distribution of the Harlequin ladybird?

....

....

(1 mark)

b Use the information to suggest **two** explanations for the change in distribution of the Harlequin ladybird.

1 ....

....

2 ....

....

(2 marks)

c The arrival of the Harlequin ladybird has put many of the native species of ladybird at risk. Use the information to suggest an explanation for this.

....

....

....

....

(3 marks)

d The scientists used hundreds of members of the public throughout the UK to collect data of sightings, backed up by photographic evidence. From this data they produced their maps.

i How has this approach led to reliable results?

....

(1 mark)

ii Why did the scientists require photographic evidence?

....

(1 mark)

(Total marks: 8)

Revision objectives

- ✓ explain that differences in characteristics (variation) may be due to differences in genes, the environment, or both
- ✓ know that most body cells contain chromosomes, which carry information in the form of genes
- ✓ know that genes control the characteristics of the body
- ✓ identify the two forms of reproduction – sexual and asexual
- ✓ know that new plants can be produced from cuttings

Student book references

1.23 Variation

1.24 Reproduction

Specification key

✓ B1.7.1, ✓ B1.7.2 a – b

Variation

Variation is the differences between individuals. These differences are not only between individuals of different species but are also between individuals of the same species. For example, we all look slightly different.

Variation can be caused by:

- **Genes** – these are inherited from our parents. Examples of **characteristics** that are controlled by genes are eye colour, earlobe shape, and flower colour.
- Environment – this is where the conditions in our surroundings influence a feature. Examples of environmental characteristics are scars, and the number of flowers produced on a rose bush being dependent on the amount of sun.
- Combination – both genes and the environment interact to determine a feature. Examples of characteristics controlled by a combination are height and body mass.

Genes and variation

Genes are major contributors to variation in organisms.

What are genes?

Genes are found in the nucleus of the cell.

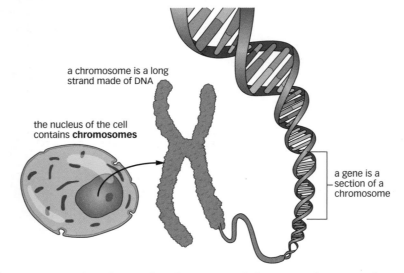

a chromosome is a long strand made of DNA

the nucleus of the cell contains **chromosomes**

a gene is a section of a chromosome

Genes are molecular codes that control the manufacture of proteins. These proteins influence the development of characteristics. For example, genes code for the pigments that colour the eyes. Different genes code for different characteristics.

Passing on genes

Genes can be passed on from generation to generation. They pass in the egg and sperm cells, which are called **gametes**. The two gametes join to form the baby.

Key words

variation, gene, chromosome, characteristic, sexual reproduction, asexual reproduction, gamete, fertilisation, cutting, clone

Reproduction

Reproduction is the production of new individuals of the same species. There are two types of reproduction.

Sexual reproduction

Sexual reproduction involves the production of sex cells or gametes by the adults. The male produces sperm cells and the female produces egg cells. During **fertilisation** the two gametes are brought together and fuse to form the offspring.

We share characteristics with our parents. This is because we have inherited half of our genes from our father in the sperm cell and half of our genes from our mother in the egg cell. This means that we are genetically different from each parent, and will be a mix of characteristics of our father and mother. Sexual reproduction leads to great variation.

Asexual reproduction

Asexual reproduction does not involve the production of gametes. This means:

* only one parent is needed
* there is no mixing of genetic information
* all offspring are genetically identical. They are called **clones**.

The advantage is that the process is quick and individuals do not need to find a mate. The disadvantage is the lack of variation. Asexual reproduction occurs in bacteria, many single-celled organisms, and plants.

Plant cuttings

Gardeners use an asexual technique to produce large numbers of plants. They take **cuttings** from a parent plant, which will grow into new plants. The advantages are that:

* the process is cheap
* the new plants are all genetically identical and so share the characteristics the gardeners want.

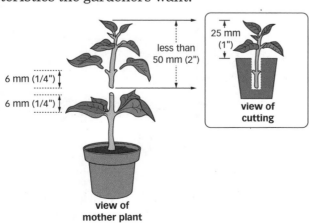

6 mm (1/4")
6 mm (1/4")
less than 50 mm (2")
25 mm (1")

view of cutting

view of mother plant

Exam tip   AQA

Learn the difference between a gene, a chromosome, and DNA; students often muddle these three words.

Questions

1 What do scientists call the differences in characteristics between individuals?

2 What causes these differences in characteristics between individuals?

3 **H** What is the difference between sexual and asexual reproduction?

Revision objectives

- ✔ know the three different modern cloning techniques
- ✔ make informed judgements about issues concerning cloning
- ✔ understand that genes can be transferred from a cell of one organism to another
- ✔ know that genes can also be transferred to the cells of animals or plants at an early stage in their development so they develop desired characteristics

Student book references

1.25 Cloning

1.26 Genetic engineering

Specification key

✔ B1.7.2

Modern cloning techniques

A **clone** is an organism that is genetically identical to its parent. These can be produced by asexual reproduction. Modern biologists can also produce clones in the laboratory. There are three common techniques.

Tissue culture

- Small groups of plant cells are taken from the parent plant, often from the shoot tip.
- They are placed on agar jelly, containing plant hormones.
- New plants start to grow.

◀ Used to produce expensive orchids.

Embryo transplants

- Parents are selected with the desired characteristics.
- Their eggs are collected and fertilised with sperm in a dish.
- The embryos are allowed to develop into a ball of unspecialised cells.
- The ball of cells is then split up into pairs of cells.
- Each pair continues to develop, and can be transplanted into host mothers called surrogates.
- They give birth to identical offspring with the desired characteristics.

Adult cell cloning

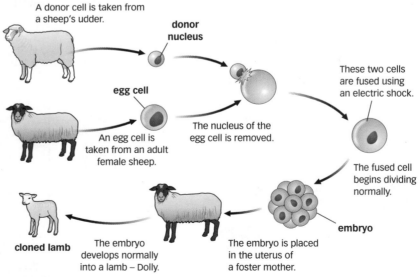

A donor cell is taken from a sheep's udder.

donor nucleus

egg cell

An egg cell is taken from an adult female sheep.

The nucleus of the egg cell is removed.

These two cells are fused using an electric shock.

The fused cell begins dividing normally.

embryo

The embryo is placed in the uterus of a foster mother.

cloned lamb

The embryo develops normally into a lamb – Dolly.

▲ Destroying embryos, lack of variation, and unknown long-term effects are ethical concerns linked to these types of animal cloning.

Genetic engineering

This is a technique where a desired gene is removed from the chromosomes of one organism (the donor) and transferred into a cell of a second organism (the host).

The host acquires the new desired characteristics, such as making new products like **insulin**. Insulin made this way is human insulin, and is better than insulin removed from pigs and cows because it is a closer match and there is no risk of disease transmission.

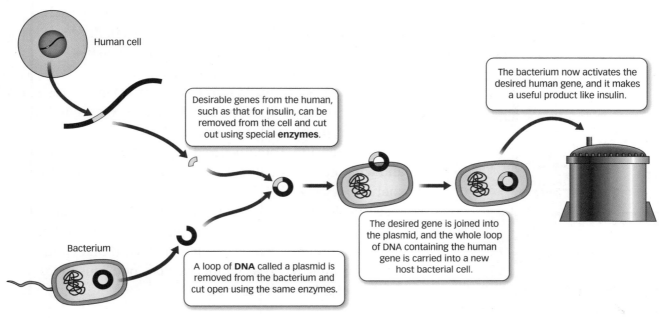

Human cell

Desirable genes from the human, such as that for insulin, can be removed from the cell and cut out using special **enzymes**.

Bacterium

A loop of **DNA** called a plasmid is removed from the bacterium and cut open using the same enzymes.

The desired gene is joined into the plasmid, and the whole loop of DNA containing the human gene is carried into a new host bacterial cell.

The bacterium now activates the desired human gene, and it makes a useful product like insulin.

▲ Genetic engineering process used to produce a desired characteristic.

Interfering with genes like this could create bacteria that could become dangerous to humans, so-called 'superbugs'. There are also ethical concerns with genetic engineering.

Questions

1   Name two methods by which scientists can produce clones.

2   Why do biologists carry out genetic engineering?

3   **H** What is genetic engineering? Suggest an ethical concern it raises.

Making a genetically modified organism

Biotechnologists do not only put genes into bacteria. They can also transfer genes into other organisms, including plants and animals, at an early stage in their development. These are then called **genetically modified** (GM) organisms. By adding new genes their genetic code is altered so that they develop with desired characteristics.

This technique is often used with agricultural organisms. It results in GM crops and animals that are of increased economic value. The characteristics improve their survival and therefore the **yield**. Examples of characteristics transferred by GM technology include:

- **Herbicide resistance** in plants. Soya is herbicide resistant, so when grown in fields the herbicide can be used to kill competing plants and the soya survives.
- Longer shelf life. Tomatoes and melons do not ripen as fast, and will not then start to decay and go soft on supermarket shelves.
- Insect resistance in plants. Corn and cotton have genes that kill insect pests, which stops the plants being destroyed.
- Increased vitamin content. Golden rice contains vitamin A in its grains, whereas white rice does not. This helps prevent blindness in children.

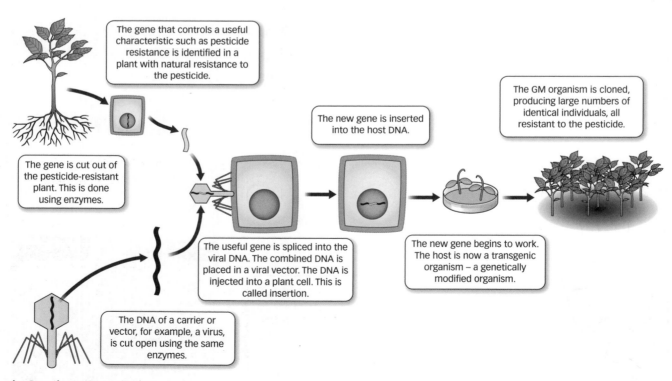

The gene that controls a useful characteristic such as pesticide resistance is identified in a plant with natural resistance to the pesticide.

The gene is cut out of the pesticide-resistant plant. This is done using enzymes.

The DNA of a carrier or vector, for example, a virus, is cut open using the same enzymes.

The useful gene is spliced into the viral DNA. The combined DNA is placed in a viral vector. The DNA is injected into a plant cell. This is called insertion.

The new gene is inserted into the host DNA.

The new gene begins to work. The host is now a transgenic organism – a genetically modified organism.

The GM organism is cloned, producing large numbers of identical individuals, all resistant to the pesticide.

▲ Creating a GM organism.

Evaluating GM crops

There are many arguments for and against the use of GM crops. Here are a few.

Arguments for GM crops	Arguments against GM crops
GM crops require fewer **pesticides** in order to grow, therefore there is less pollution in the environment.	GM crops might escape from the farms and become more successful competitors than wild flowers in the environment (superweeds).
GM crops have a higher yield and can feed larger populations. This is important in developing countries.	It disrupts the food chain if insect-resistant plants are grown, killing the insects.
GM foods have been eaten in some countries for 10 years with no ill effects noted.	There is uncertainty about the effects of eating GM crops on human health.
Some GM crops can have a higher nutritional value, for example, golden rice has a higher vitamin A content.	GM crops are expensive to develop and test. Seeds may be expensive.
The process produces crops with desired characteristics far quicker than using selective breeding.	GM crops require a lot of testing, which can take years.

Some countries are so concerned about these issues that they do not grow GM crops. There are no commercial GM crops in the UK, for example. Other countries, such as the USA and countries with greater needs for larger crops, such as India, are less concerned.

Key words

genetically modified, herbicide, resistance, yield, pesticide

Exam tip

When asked to make judgements about the pros and cons of techniques like cloning and genetic modification, use scientific arguments to back up your viewpoint.

Questions

1 What are GM crops?

2 What advantages are there to GM crops?

3 **H** Discuss an ethical concern relating to GM crops.

Revision objectives

- ✓ know that classification groups organisms based on similarities and differences
- ✓ know the causes and effects of evolution
- ✓ know that evolution can result in the formation of new species
- ✓ know that Charles Darwin produced the theory of evolution, but that other scientists have produced other explanations

Student book references

1.29 Classification

1.30 Surviving change

1.31 Evolution in action

1.32 Evolutionary theory

Specification key

✓ B1.8.1

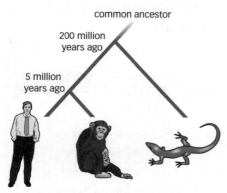

▲ This family tree shows the relationship between three species.

common ancestor

200 million years ago

5 million years ago

Classification

Biologists group living organisms based on their **similarities** and differences. This is called **classification**. These studies show two types of link.

Ecological links

Here organisms may share features because they live in the same environment. For example, butterflies and birds both have wings because they fly, but the wings are very different in structure.

Evolutionary links

Here organisms with a common ancestor may share characteristics. For example, humans, chimpanzees, and lizards have limbs with digits. These limbs are similar structures, as the animals have a distant common ancestor. The more similarities in two groups of organisms, the more closely they are related. Chimpanzees and humans are more closely related than lizards and humans as their limbs are more similar.

All living things are placed first into large groups called **kingdoms**. There are three kingdoms: plant, animal, and microorganism.

Evolution

Biologists needed to be able to explain how the great variety of different **species** had developed. Did they change over time? Charles **Darwin** suggested a theory that could explain how all of these species might have developed or evolved from simple life forms that first appeared over three billion years ago. His theory was called **evolution** by **natural selection**.
Evolution is the gradual change of an organism over time. His idea suggests that gradually one type of organism, called an ancestor, might change over many generations into one or more different species. This would generate all of the different species in the world over millions of years.

The theory was only gradually accepted because:
- the theory disagreed with most religious ideas about how God created all life forms
- there was not much evidence at the time to convince other scientists
- scientists did not know about genes and how they are inherited, and cause variation, until 50 years after the theory was published, and so could not explain a mechanism for how evolution could work.

Individuals in a large population of a species show variation. This is because of differences in their genes. More variation is caused by chance **mutations** in the genes.

Within any population there is always a struggle to survive because of a number of factors such as predators, shortage of food, or disease.

Some variations are more suited to the environment. They are the 'fittest', and are more likely to survive and breed.

The surviving individuals are the only ones that reproduce, passing on their genes to the next generation.

Generally evolution is quite slow but mutations cause rapid changes to an individual in the population. If this is combined with a change in the environment favouring the mutation then the evolution of change in the species as a whole is much more rapid.

Forming a new species

Where evolution continues to occur in a population of a species, there may be many changes over time. This might result in the formation of a completely new species. A good example of this is seen in the elephant population. The African and Indian elephants are separate species that have both evolved from a common ancestor over time.

Alternative theories

Charles Darwin was not the only biologist to attempt to explain the origin of all the species on Earth. Other theories have been suggested, including that of Jean Baptiste **Lamarck**. These theories often suggest that the changes occur in an individual of a species during its life, making it better suited to its environment, and that these changes would then be passed on. However, we now know that genes control our characteristics and that they do not change during an organism's lifetime.

Key words

classification, kingdom, similarities, evolution, natural selection, mutation, species, Darwin, Lamarck

Exam tip

Learn the four key steps in the process of natural selection, and be able to apply them to any organism given in a question.

Questions

1   Who was Charles Darwin?

2   What is natural selection?

3   **H** What is the key difference between Darwin's explanation of evolution and that of Lamarck?

1 What is variation?

2 Variation is caused by several factors. Which factor is most likely to have caused:
  a the colour of a flower?
  b the height of a human?
  c the colour of the fur on a dog?
  d how much a horse weighs?

3 Which chemical is a chromosome made of?

4 Where are chromosomes found in a cell?

5 What types of cell are involved in sexual reproduction?

6 What is fertilisation?

7 Name two types of organism that carry out asexual reproduction.

8 What is a clone?

9 Name one product we get from genetic engineering.

10 What is used to cut DNA?

11 Give one way in which corn has been genetically modified.

12 Do all countries grow GM crops?

13 Give two arguments people use against GM crops.

14 What is a mutation?

15 Who proposed the theory of evolution by natural selection?

16 What are the three main kingdoms?

17 Place the following organisms into one of the three kingdoms:
  a oak tree
  b human
  c fish
  d daffodil
  e *E. coli*
  f crab
  g *salmonella*

18 Look at the drawing below.

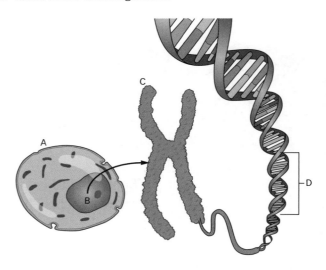

  a Which letter labels the gene?
  b What do genes control?

19 Describe how genes are passed on from one generation to the next.

20 Explain why we share characteristics with our parents.

21 Explain why we are not identical to our parents.

22 What is the advantage to the gardener of using cuttings of plants for reproduction?

23 During tissue culture a number of steps occur. Put the following steps into the correct order:
  a The jelly contains special chemicals that cause roots to grow.
  b Small groups of cells are taken from a plant.
  c A second jelly contains chemicals that cause stems and leaves to grow.
  d Cells are placed in a liquid or jelly.
  e A new plant has formed, identical to the original.

24 What is a surrogate?

25 Explain why scientists carry out embryo transplants in animals.

26 Name one type of mammal that has been cloned.

27 How do biologists make an egg cell and a donor nucleus fuse?

28 Look at the diagram below, which outlines the process of genetic engineering.

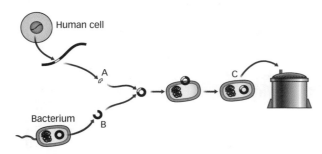

a What letter on the diagram shows the following?
   i plasmid
   ii desirable gene
   iii bacterium
b Why is insulin made by bacteria of more use to humans than insulin obtained from pigs?

29 Soya plants have been genetically modified to be resistant to weedkiller. How will this improve the yield of soya?

30 Why is genetic engineering preferred over selective breeding?

31 There are many arguments for the use of GM crops.
a Which argument would also help protect the environment?
b Which argument would suggest that our health concerns are not valid?
c Which argument shows that GM crops can help poorer countries who struggle to feed their populations?

32 How do scientists classify organisms?

33 In evolutionary terms, why are the limbs of humans and chimpanzees so similar?

34 Why will a biologist place humans and dogs as closer relatives than humans and fish?

35 Why do sharks and dolphins have similar streamlined bodies when they are not closely related?

36 Give two reasons why Darwin's theory was not initially accepted.

37 Why do biologists think that the explanation for evolution made by Lamarck is no longer acceptable?

38 How long ago do scientists think that life arose on Earth?

39 Explain how a gene would determine the colour of your eyes.

40 Explain why humans need thousands of genes.

41 What are the major differences between sexual and asexual reproduction?

42 Outline the advantages and disadvantages of asexual reproduction.

43 Explain why taking cuttings of plants is a method of asexual reproduction.

44 Suggest one beneficial outcome that could come from cloning.

45 How would the process shown in question 28 be altered to insert a gene into a sheep?

46 A common moth in the UK is the peppered moth. It exists in two main forms, the light form and the dark form.

light moths

dark moths

light tree   dark tree

The light moth was once the only type of peppered moth. It was widespread over the whole of the UK, but now survives well in the cleaner areas of the UK, like Cornwall, whilst the darker moth developed, and survives better, in areas with heavier air pollution, like the industrial Midlands.
a Use Darwin's ideas of natural selection to explain how the darker variety became common in the industrial areas of the UK.
b Suggest why the dark moth would not survive well in the cleaner areas of the UK.
c What caused the dark moth to appear?
d How might Lamarck have explained the development of the darker moth?
e Why did the evolution of the moth happen very quickly?

1   Look at the diagram below, which shows a technique that scientists use to produce many plants from a valuable specimen.

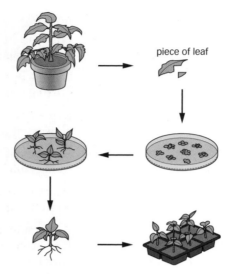

piece of leaf

a   All the plants produced are identical to the parent. What is the name given to these types of organism?

....

(1 mark)

b   Describe how the sections of leaf are encouraged to produce roots and shoots.

....

....

....

(2 marks)

c   Explain why gardeners prefer to use this technique to produce plants, instead of allowing the plants to reproduce sexually.

....

....

....

(2 marks)
(Total marks: 5)

2 When biologists began to study the animals of South America, they discovered the anteater. Anteaters are related to the sloths. They have developed a tubelike snout with a long tongue to eat termites and ants from mounds.

Use your knowledge of the theory of evolution to explain as fully as you can how the anteater evolved the long snout.

....

....

....

....

....

....

....

....

....

....

(4 marks)
(Total marks: 4)

3 Read the following passage about the production of GM crops.

> **GM crops – growing tomorrow's food**
>
> The development of GM crops was seen by many as the solution to world hunger. Biologists are able to produce crops that are pest-resistant, pesticide-resistant, and slow to decay. In addition, crops can even be created that have increased nutritional value, such as golden rice.
>
> There are some in our society who have voiced concerns about such crops. The fear is that genes could escape from the GM crops, creating 'superweeds'. They also wonder if the GM food would cause health problems, such as allergies, to consumers.
>
> The production of such crops is expensive and some government economists think poorer countries could not afford them anyway.

Use the information and your own knowledge and understanding of biology to evaluate whether the production of GM crops is such a great benefit to the world population. Remember to give a conclusion to your evaluation.

....

....

....

....

....

....

....

....

....

(5 marks)
(Total marks: 5)

Designing investigations and using data

Within this module there are some very good examples of the design of investigations and interpretation of data. You may be asked in a question to demonstrate these skills yourself.

Look at the example below; this will show you how scientists design experiments and interpret data. There will be opportunities for you both to practise some of the skills, and to be guided through some of the more difficult elements that might appear in a question in the exam.

Pollution and indicator species

Skill – designing investigations

> Biologists were concerned about the levels of pollution being released into a river by a factory, and the impact the pollutants might have on the organisms that live in the river.
>
>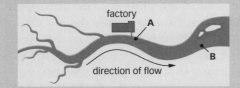
>
> Biologists set out to investigate the levels of pollution in the river. At point A on the map (where the factory was) they recorded two variables:
> * the oxygen levels in the water, using a digital oxygen probe
> * the number of rat-tailed maggots in a square metre.
>
> 1 How did the biologists ensure that the readings for the oxygen levels were:
>   a accurate?
>   b reliable?

These questions seem simple, but always cause trouble. Students often talk about the need to keep the experiment 'fair' here. That is the *reason* for doing these things, but not *how* these points are achieved.

You need to remember a simple rule:
For **ACCURACY** use the right **APPARATUS**.
For **RELIABILITY** do **REPEATS**.

> The biologists repeated the experiment at point B (100 metres downstream of the factory). Not only was the data for the maggots and oxygen recorded but also several other possible pollutants. The data is in the table in the next column.

Pollutant and indicator species	Point A At the factory	Point B 100 metres from factory
Oxygen	3.6 mgO$_2$/l	9.8 mgO$_2$/l
Nitrates	20.4 mgNO$_3$/l	7.8 mgNO$_3$/l
Calcium	40 ppm	39 ppm
Magnesium	24 ppm	26 ppm
Zinc	0.02 mg/l	0.02 mg/l
Lead	0.02 ppm	0.01 ppm
Total invertebrate species	14	103
Number of rat-tailed maggots	36	2

> 2 What does the distribution of the rat-tailed maggots tell us about the levels of pollution in the river?

This question is asking you to look at the data in the table.
* You must identify a specific piece of data, and then interpret that data.
* Do not get over-concerned with the other pieces of data when answering this question.
* Look at the data: it is telling us that there are more maggots in the water by the factory, which is probably polluted.
* The question then needs a link to your knowledge of the distribution of maggots.
* It is asking you to discuss the relationship between two variables: the number of maggots and the amount of pollution.

> 3 The factory is responsible for the release of the pollutants into the water.
>   a How can biologists tell that the pollutants caused the growth of respiring bacteria?
>   b What was the major pollutant released by the factory?

In both of these questions you are being asked to select appropriate information from the table. Remember:
* Look for pieces of data that show some significant difference or change.
* In this table only the oxygen and nitrate levels show change.
* Think about whether those pieces of data are biologically relevant to either of the questions. In this case, the oxygen levels link to the question about respiring bacteria, and the nitrates relate to the pollutants.
* Not all of the data is needed, so much of it might be ignored.

Answering extended writing questions

Giraffes have evolved to have long necks.

1 Charles Darwin proposed a theory to explain how animals evolved. What is the name of this theory? *(1 mark)*

2 How would Charles Darwin have explained the way the giraffe had evolved a long neck? *(4 marks)*

QUESTION

G–E

1 Evolution.
2 A giraffe developed a long neck to adapt to its environment. This adaption allows the animal to survive betterer in its environment. This is cause it can reach on top of the tree to get more food. Because it can get the leaves at the top of the tree it can survive better than the other giraffes which cannot get the leaves at the top of the tree. It now will survive and the short one wont. So all the giraffes now grow long necks.

Examiner: This student has either not understood the question or not learnt the correct name.

This answer is worth one mark at best. It only really gets the idea of the longer necks giving an advantage. There are a number of problems. The answer is rambling, and there is considerable repetition, so it is not logical. There are some spelling errors, and little use of technical terms. Adaption is an American term. The student appears to take the question into the wrong topic, that of adaptation. Also one major issue is that the student says that a giraffe develops a long neck, which suggests that the giraffe alters its own body to improve its survival, whereas evolution is about the random development of mutations that might give advantage.

D–C

1 Survival of the fittest.
2 Some giraffes had long necks, and they were able to eat the leaves at the tops of the trees. This gave them an advantage over the short-necked giraffes. This meant that they could have lots more food. These animals were the fittest. They grew more than the shorter ones who couldn't get food. Eventually all the giraffes had longer necks, the shorter ones died out.

Examiner: This is a shame. The candidate has clearly understood some of the ideas of evolution, but has muddled the name of the theory with the way it works.

This answer is worth two marks. Here the candidate has some understanding of evolution. However, they have not used many technical terms; for example, there is no mention of genes or mutations. Also, they have not always explained each idea clearly. For example, they have not really said what survival of the fittest is, they have just used the term. This will fail to gain them credit for these points. The answer is not a complete account of the process, as they have not mentioned reproduction and passing on characteristics at all.

B–A*

1 Natural selection.
2 In Africa there were large numbers of giraffes. They were slightly different: some had short necks, but some had grown a long neck. The long neck was caused by a mutation in one of their genes. The giraffes all competed with each other for limited food resources. The giraffes with the longer necks had an advantage in that they could reach leaves at the tops of trees, and so they survived. They could then reproduce more and pass on their genes. Gradually only the giraffes with long necks survived.

Examiner: Well answered. Candidate has learnt the correct name.

A complete and thorough answer. Full marks awarded. It makes seven clear points, far more than needed by the mark scheme. This is good practice, because if one of the statements had been poorly expressed and not worthy of the mark, there are other points that will pick up the marks. This raises a good point – in longer-answer questions there are usually more marks possible on the mark scheme than you need. This gives you room to say different things. The answer given here is also logical and well expressed. No spelling errors.

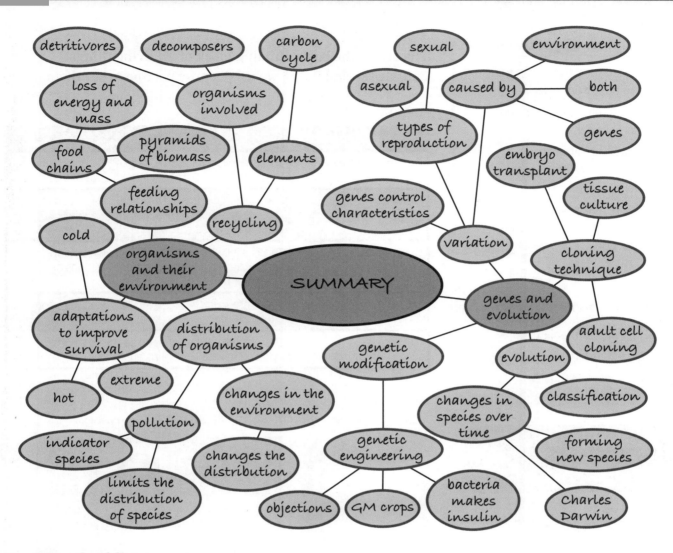

Revision checklist

○ Living things show adaptations that help them to survive in their environment.

○ Organisms show adaptations to hot, dry, and cold environments. Some are even adapted to survive in extreme environments.

○ If the environment changes it can affect the distribution of organisms.

○ Human pollution affects the environment and the organisms that live there. Indicator species are organisms that show how polluted an area is.

○ Food chains show the flow of biomass and energy from one organism to the next. The mass reduces as you move along a food chain or up a pyramid of biomass.

○ Energy is lost at each step of the food chain, as heat and in waste materials.

○ Elements are constantly cycled between the living and non-living world.

○ The carbon cycle shows how carbon moves into the living world by photosynthesis, and is released back into the non-living world by respiration and burning.

○ Variation is the difference in characteristics shown between organisms of the same species. It is caused by genes, the environment, or a combination of both.

○ Reproduction can be sexual, using sex cells to create variation, or asexual, with no sex cells, producing identical clones.

○ Clones can be produced by tissue culture, embryo transplant, and adult cell cloning.

○ Genetic engineering is the transfer of a useful gene from one organism to another. This allows us to transfer a useful characteristic. Using this method we can produce drugs like insulin.

○ GM crops are crops that have been given genes that provide an advantage to them, for example, a higher yield. However, some scientists have concerns about the technique.

○ Classification is the process where organisms are grouped together based on their similarities and differences.

○ Evolution is the gradual change of organisms over time. Charles Darwin suggested a theory called natural selection to explain how this happened, but it took years to be accepted.

Revision objectives

- ✓ understand that living things are built from cells
- ✓ know the function of the parts of cells
- ✓ know the differences between different types of cells
- ✓ know that cells can become specialised for a specific function

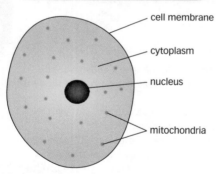

▲ A typical animal cell.

Exam tip AQA

In this topic the key point is to know the parts of the cell, and what they do. Students often muddle them. Practise labelling cell parts on diagrams of the different types of cell, and give a function for each part.

The cell

The **cell** is the basic building block of all living things. Cells are so small that we need a **microscope** to see them.

Animal cell

Name of cell part	Function
Cell membrane	This is a thin layer around the cell. It controls the movement of substances into and out of the cell.
Nucleus	This is a large structure inside the cell. It contains chromosomes, which control the activities of the cell, and how it develops.
Cytoplasm	This is a jelly-like substance containing many chemicals. Most of the chemical reactions of the cell occur here.
Mitochondria	These are small rod-shaped structures that release energy from sugar during aerobic respiration.
Ribosomes	These are small ball-shaped structures in the cytoplasm, where proteins are made.

Plant cell

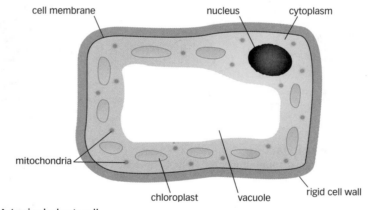

▲ A typical plant cell.

Plants have all of the structures of animal cells, plus a few others.

Part of cell	Function
Cell wall	This is outside the cell membrane. It is made of cellulose, which is strong and supports the cell.
Permanent vacuole	This is a fluid-filled cavity. The liquid inside is called cell sap, used for support.
Chloroplasts	These are small discs found in the cytoplasm. They contain the green pigment chlorophyll. Chlorophyll traps light energy for photosynthesis.

Other types of cell

Other groups of organisms are also made of cells, although the cells might have slight differences from plant or animal cells.

Algal cells

Algae are an important group of organisms, which includes the seaweeds. Their cell structure is the same as plant cells.

Bacterial cells

In **bacterial** cells, many of the structures found are similar to those of plants or animals, like the cytoplasm, membranes, and ribosomes. But:

* the cell wall has a similar function to a plant's, but it is made of a different chemical
* there is no distinct nucleus but the cell does have DNA, which is in the form of a loop.

Fungal cells

Again, the structures of the fungal cell are similar to those of plants and animals. They have a nucleus, cytoplasm, membranes, and cell wall.

Fungi are larger than bacteria and include important examples like the single-celled yeasts used in bread- and beer-making.

Special cells for special jobs

Although cells in a human or a plant have the same basic structures, they often carry out different jobs or functions. Cells become **specialised** to carry out a particular job, by developing special structures. This is called **differentiation**.

For example:

* Red blood cells – these lack most cell structures but contain large amounts of haemoglobin, which allows them to carry oxygen.
* Muscle cells – these contain contractile proteins, which allow the cell to shorten.
* Palisade cells – these contain many chloroplasts, which allow the cell to photosynthesise.
* Root hair cells – these have a long extension, which projects into the soil and increases the surface area for absorbing water.

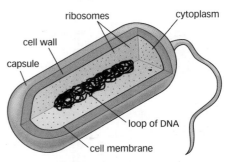

▲ A typical bacterial cell.

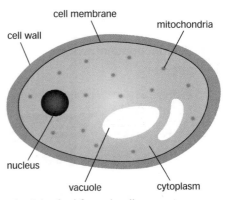

▲ A typical fungal cell.

Questions

1. What is the nucleus?
2. How could you identify a plant cell and an animal cell?
3. **H** What is differentiation?

Exam tip AQA

Students don't usually find the idea of diffusion difficult, but they tend to make three mistakes. Remember: **1** the movement of particles is random; **2** it is the net movement; **3** the movement is from high concentration to low.

Questions

1 Why is diffusion important for living things?

2 Why can't diffusion occur in solids?

3 What is the relationship between surface area and the rate of diffusion?

Movement of molecules

It is important for molecules to be able to move into and out of cells for them to work. **Diffusion** is one important method for achieving this.

Diffusion

Diffusion is the net movement of particles from an area of high concentration to an area of low concentration, until the concentration evens out. This happens in a liquid or gas where the particles can move.

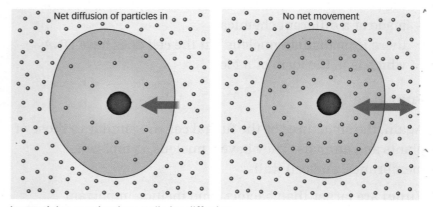

▲ Particles moving into cells by diffusion.

Examples of diffusion are:
- Oxygen diffuses into cells for use in respiration.
- Carbon dioxide diffuses out of cells as the waste product of respiration.

Factors affecting the rate of diffusion

Particles are constantly moving. Diffusion is the net movement of particles in one direction. Several factors affect the rate at which particles move:
- Distance – the shorter the distance the particles have to move, the quicker the rate of diffusion. For example, leaves are thin so carbon dioxide can move through the leaf quickly.
- **Concentration gradient** – particles move down a concentration gradient from high to low concentration. The greater the difference in concentration, the faster the rate of diffusion.
- Surface area – the greater the surface area means that there is more surface over which the molecules can move, so the rate is faster. For example, the lungs have a large surface area for the movement of oxygen.

Working to Grade E

1 Look at this diagram of an animal cell.

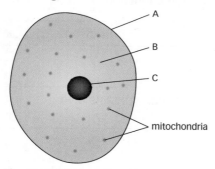

a Which of the labels A to C are:
  i the cell membrane
  ii the nucleus
  iii the cytoplasm?
b List **three** structures you would expect to find in a plant cell that are not present in the animal cell.
2 Where is the cell sap found in a plant cell?

Working to Grade C

3 What is the function of:
  a the nucleus
  b the cell wall
  c the cell membrane
  d the chloroplast?
4 Look at these drawings of three cells found in either plants or animals.

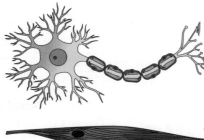

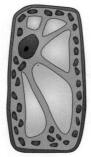

For each of the cells:
a identify a special part of the cell
b explain how this part helps the cell perform its function.

5 What is the name of the process where cells become specialised?
6 Draw and label a yeast cell.
7 Why are yeasts unable to photosynthesise?
8 Define diffusion.
9 Give **one** example of a molecule that moves into a cell by diffusion.
10 List **three** factors that might affect the rate of diffusion.

Working to Grade A*

11 Ribosomes are found in all cell types. What do they do?
12 Describe the differences between a bacterial cell and a plant cell.
13 Look at this diagram of a cell. There are different chemicals at each corner labelled A to D.

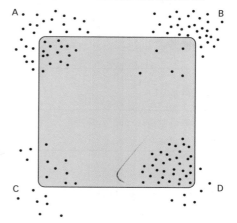

Indicate on the diagram whether diffusion is occurring at each corner, and if so in which direction it will occur.
14 Explain what is meant by a concentration gradient.
15 Diffusion occurs in cells.
  a Explain what is meant by the rate of diffusion.
  b Explain how the three factors from question 10, above, would affect the rate of diffusion.
16 Diffusion occurs efficiently in the lungs. Explain why this is so.

1 Look at the table.

a Complete the table by filling in the name of either the structure or the function.

Structure	Function
Cell membrane	*to allow substances to enter and leave cell*
nucleus	Controls the cell, contains DNA.
Cell wall	
Mitochondria	Releases energy in aerobic respiration.
Chloroplasts	
Vacuole	
	Where many chemical reactions occur.
ribosome	Proteins are made here.

(8 marks)

b Of the structures listed above, identify **three** that are only present in plant cells.

1 *chloroplast*

2 *vacuole*

3 *cell wall*

(3 marks)
(Total marks: 11)

2 Diffusion occurs in cells.

a Define diffusion.

diffusion is the movement ... area of high concentration ... to low concentration

(2 marks)

b Name a gas that would diffuse into a cell for use in respiration.

(1 mark)
(Total marks: 3)

Working together

Animals and plants are **multicellular**, which means built of many **cells**. Cells do not work in isolation. The cells in our body are organised.

Cell for example, muscle cell	The basic building block, which can become specialised to perform a particular function.
↓	
Tissue for example, heart muscle	A tissue is a group of similar cells working together.
↓	
Organ for example, heart	An organ is a group of different tissues working together at a specific function.
↓	
Organ system for example, circulation	An organ system is a group of different organs working together at a specific function.

Revision objectives

- understand the levels of organisation of cells, tissues, and organs
- appreciate the working of some animal tissues and organs
- know that organs work together in organ systems
- know the major organs of the digestive system

Student book references

2.5 Animal tissues and organs

2.6 Animal organ systems

Specification key

- B2.2.1

Animal tissues

There are many examples of tissues in animals.

Muscle tissues – these are able to contract to bring about movement.

Glandular tissues – these produce substances like enzymes and hormones.

Epithelial tissues – these tissues act as covering for parts of the body.

Animal organs

A good example of where tissues work together in an organ is the stomach.

Epithelial tissue covers the outside of the stomach.

Muscular tissue, which can contract causing the wall of the stomach to move. This will churn the contents up.

Glandular tissue, which produces acid and enzymes that are poured into the stomach cavity to help digest food.

Inner epithelial tissue to cover the inside of the stomach.

▲ The stomach is an organ that has several different types of tissue working together.

Animal organ systems

Organs work together to form organ systems. An example of an organ system in the body is the **digestive system**. This is an exchange system that has two major functions in the body:

- Digestion – where food is broken down. Juices, containing enzymes, produced in glands are released into the gut to digest the food.
- Absorption – where useful molecules are taken from the gut into the blood.

The different organs include:

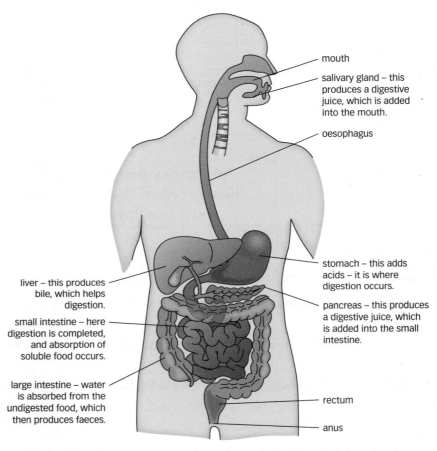

mouth

salivary gland – this produces a digestive juice, which is added into the mouth.

oesophagus

stomach – this adds acids – it is where digestion occurs.

pancreas – this produces a digestive juice, which is added into the small intestine.

liver – this produces bile, which helps digestion.

small intestine – here digestion is completed, and absorption of soluble food occurs.

large intestine – water is absorbed from the undigested food, which then produces faeces.

rectum

anus

▲ In the digestive system several organs work together to bring about digestion.

Exam tip   AQA

You should be able to recognise and label the organs of the human digestive system. Practise labelling diagrams.

Questions

1   What does multicellular mean?

2   Name **three** organs in the human digestive system.

3   H What is the difference between an organ and a tissue?

Plant organs

Plants are also organised into tissues, organs, and organ systems.

Plant organs include:

Organ	Function
Stem	Supports the plant. Transports molecules through the plant.
Leaf	Produces food by photosynthesis.
Root	Anchors the plant. Takes up water and minerals from the soil.

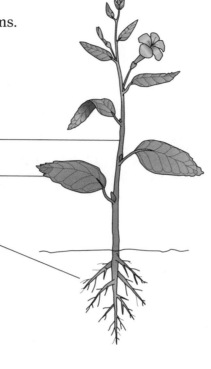

Plant tissues

The organs of the plant are made from tissues, just as in an animal. Examples of plant tissues include:

- **epidermal tissues**, which form a covering layer over the surface of the plant
- a **mesophyll** tissue layer inside the leaf, which contains cells loaded with chloroplasts. These cells carry out photosynthesis.
- **xylem** tissues, which are made of hollow cells with strong cell walls. The cells are stacked one above the other and form a long tube through the plant. Xylem tissue is found around the edge of the stem. These cells carry water from the roots to the leaves, and help support the plant.
- **phloem** tissues found close to the xylem. Again phloem form long tubes through the plant. These cells transport sugars from the leaves to other parts of the plant.

Questions

1   Name **three** tissues of a plant.

2   Which tissues are involved in transport in plants?

3   **H** How is xylem adapted to carry out the function of transporting water?

Revision objectives

- ✔ know that plant cells are organised into tissues
- ✔ know the main organs of the plant
- ✔ understand the distribution and functions of some of the key tissues of the plant

Student book references

2.7   Plant tissues and organs

Specification key

✔ B2.2.2

Key words

stem, leaf, root, epidermal tissue, mesophyll, xylem, phloem

Exam tip   AQA

You need to be able to label the organs of a plant. You do not need to know the internal structure of these organs, except the leaf, which is detailed with photosynthesis.

Working to Grade E

1 Define a tissue.
2 Give an example of:
  a a tissue
  b an organ system.
3 Look at the drawings of the three human organ systems below.

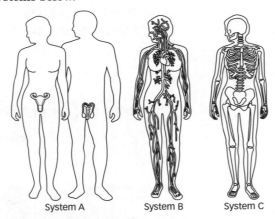

System A   System B   System C

  a Name each organ system.
  b Name an organ found in system B.
4 Define an organ.
5 Digestion occurs in the digestive system. Where do the following events occur in the digestive system?
  a Food enters the system.
  b Water is absorbed from undigested food.
  c Digestion is completed and absorption occurs.
6 Below is a diagram of the digestive system. Label parts A to H.

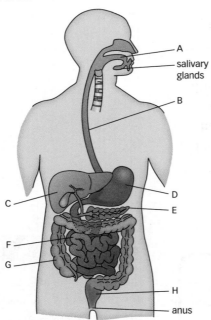

A

salivary glands

B

C

D

E

F

G

H

anus

7 What is the function of the following plant organs:
  a stem
  b leaf
  c root

Working to Grade C

8 **a** What is the function of each organ system in question 3?
  b Why are they all called organ systems?
9 The stomach is an organ.
  a What is the role of the muscular tissue in the stomach?
  b Which type of tissue produces acid and enzymes?
  c Where would you find epithelial tissues in the stomach?
10 What happens in the following organs to help digestion?
  a the pancreas
  b the liver
  c the salivary glands.
11 Why is the flower regarded as an organ system?
12 What type of tissue covers the outside of roots, stems, and leaves?
13 What does the xylem tissue transport?
14 Draw a diagram of a stem to show where the xylem and phloem would be found.
15 How are the cells arranged in the xylem tissue?
16 Where is the palisade mesophyll tissue found?
17 What is the function of the palisade mesophyll tissue?
18 What substance is transported in the phloem tissue?

Working to Grade A*

19 Explain why three different types of tissue are required in the stomach.
20 Phloem is a tissue that transports sugars.
  a Where are these sugars picked up by the phloem?
  b Where does the phloem take the sugars to?
21 There are many tissues in plants.
  a Which tissue in the plant is involved in support?
  b How are the cells of this tissue adapted for support?

1 The stomach is an organ made of several tissues.

Below is a diagram of the stomach labelled with some of the tissues that make up the stomach.

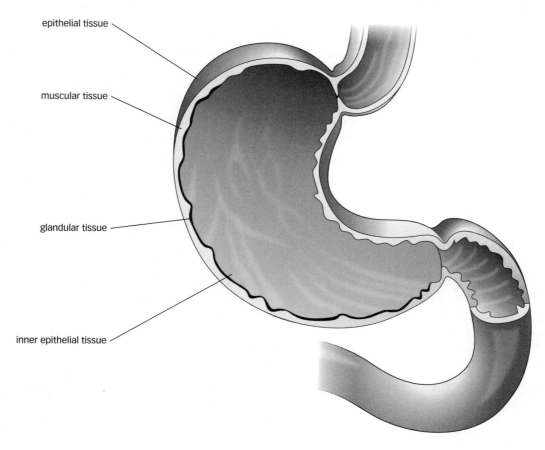

epithelial tissue

muscular tissue

glandular tissue

inner epithelial tissue

a Explain why three tissue types are needed for the correct functioning of the stomach.

....

....

....

....

....

....

....

....

....

....

(5 marks)
(Total marks: 5)

Revision objectives

- know that photosynthesis is the process by which plants make food
- know the sources of the raw materials for photosynthesis
- understand the fate of the products of photosynthesis
- appreciate the structure of the leaf as the site of photosynthesis

Student book references

2.7 Plant tissues and organs

2.8 Photosynthesis

2.9 The leaf and photosynthesis

Specification key

- B2.3.1 a – b, e – g

The process of photosynthesis

Photosynthesis is the process where plants make their food. They need two raw materials:

- carbon dioxide from the air
- water absorbed by the roots.

They also need:

- light energy from the Sun
- **chlorophyll**, a green pigment found in the chloroplasts in the cells of plants and algae, which absorbs the light.

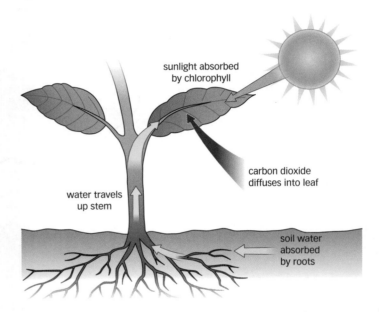

▲ In photosynthesis the plant uses sunlight energy to convert water and carbon dioxide into carbohydrates.

The equation for photosynthesis is:

$$\text{carbon dioxide} + \text{water} \xrightarrow[\text{chlorophyll}]{\text{sunlight}} \text{glucose} + \text{oxygen}$$

The light energy is needed to cause the reaction to happen between carbon dioxide and water, to make the carbohydrate glucose.

The products of photosynthesis

There are two products of photosynthesis:

- The main product of photosynthesis is the carbohydrate glucose.
- Oxygen is a by-product of the process and is released.

Glucose is the product that the plant needs. Some of it is used in respiration by the plant, but the rest is converted into other substances.

The conversion of glucose

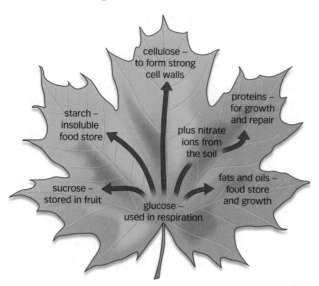

▲ Glucose from photosynthesis is converted into all the substances that a plant needs.

The leaf as the site of photosynthesis

The **leaf** is the main site of photosynthesis. It is well adapted for this function.

- The **mesophyll** contains the cells that carry out photosynthesis.
- The epidermis forms a covering layer.
- The **xylem** brings water to the mesophyll cells.
- The **phloem** takes the glucose away.
- There are **stomata** on the lower surface to allow carbon dioxide in and oxygen out.

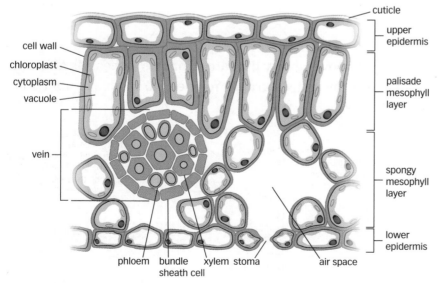

▲ The internal structure of a leaf.

Key words

photosynthesis, chlorophyll, leaf, mesophyll, xylem, phloem, stomata

Exam tip
AQA

Learn the equation for photosynthesis. Most of the questions on photosynthesis tend to ask about what is used, and what is made. You can answer these questions by knowing the equation.

Questions

1. What is the point of photosynthesis?

2. What is the energy change in photosynthesis?

3. **H** Describe how the two raw materials of photosynthesis get to the mesophyll cells in the leaf.

Revision objectives

- ✔ know how some factors affect the rate of photosynthesis
- ✔ understand limiting factors
- ✔ understand that some factors needed for photosynthesis can be controlled
- ✔ appreciate the commercial benefits of controlling photosynthesis in greenhouses

Student book references

2.10 Rates of photosynthesis

2.11 Controlling photosynthesis

Specification key

✔ B2.3.1 c – d

The rate of photosynthesis

The **rate of photosynthesis** is the speed at which a plant photosynthesises. Biologists can measure this in one of two ways:

- the amount of raw materials used up in a period of time
- the amount of product made in a period of time.

Limiting factors

When a process is affected by several factors, the one that is at the lowest level will be the factor that limits the rate of reaction. This is called the **limiting factor**.

There are three factors that limit the rate of photosynthesis:

- Availability of **light** – the less light there is, the slower the rate of photosynthesis.
- A suitable **temperature** – temperature affects the enzyme reactions. As the temperature increases so does the rate, but if the temperature becomes too high it will damage the enzymes and stop photosynthesis.
- The amount of **carbon dioxide** – the less carbon dioxide, the slower the rate of photosynthesis.

If the limiting factor is increased, then the rate of photosynthesis will increase, until one of the other factors becomes limiting.

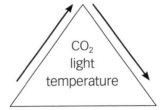

INCREASING the factor INCREASES the rate

CO_2 light temperature

DECREASING the factor DECREASES the rate

One way of showing how the rate of photosynthesis is affected by a factor is by using a graph. You need to be able to interpret these graphs.

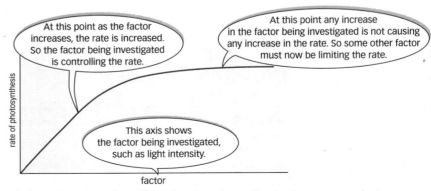

At this point as the factor increases, the rate is increased. So the factor being investigated is controlling the rate.

At this point any increase in the factor being investigated is not causing any increase in the rate. So some other factor must now be limiting the rate.

This axis shows the factor being investigated, such as light intensity.

rate of photosynthesis

factor

▲ Graph to show how the rate of photosynthesis changes as a factor increases.

Controlling photosynthesis

Gardeners and plant growers use **greenhouses** to be able to grow crops all year round, and to grow tropical plants. This is because we can control the environment inside the greenhouse to maximise the rate of photosynthesis.

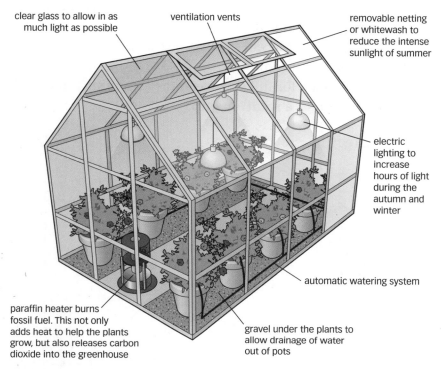

clear glass to allow in as much light as possible

ventilation vents

removable netting or whitewash to reduce the intense sunlight of summer

electric lighting to increase hours of light during the autumn and winter

automatic watering system

paraffin heater burns fossil fuel. This not only adds heat to help the plants grow, but also releases carbon dioxide into the greenhouse

gravel under the plants to allow drainage of water out of pots

▲ Modern greenhouses use automated systems to give the best conditions for photosynthesis.

The gains and costs of greenhouses

Greenhouses are used to allow plants to be grown all year round.

Factor controlled	Advantages	Disadvantages
Light	Increasing light increases the rate of photosynthesis, increasing the yield, giving greater profits.	Cost of electric lighting. Cost of nets.
Temperature	Warmth increases the rate of photosynthesis, especially during cold months. So plants can be grown out of season when they have greater value.	Glass is expensive. Cost of fuel.
Carbon dioxide	Adding carbon dioxide from burning fuels speeds up photosynthesis, increasing yield.	Cost of fuels.
Water	The correct amount of water stops plants dying and lost income.	Cost of electricity to run automated systems.

Questions

1   Why is photosynthesis faster on a sunny day?

2   How do gardeners use greenhouses to alter the rate of photosynthesis?

3   **H** What changes should be made to the limiting factors to speed up plant growth?

Living in the environment

Plants and animals live in many environments on Earth. Biologists need to make sense of where things live.

- They note where a species lives – this is the **distribution**.
- They count the number of individuals of a species in an area – this is the **population**.
- They look at the different populations that live together in an area – this is the **community**.
- Finally they look for links between the community and external factors – a **relationship**.

Factors affecting distributions

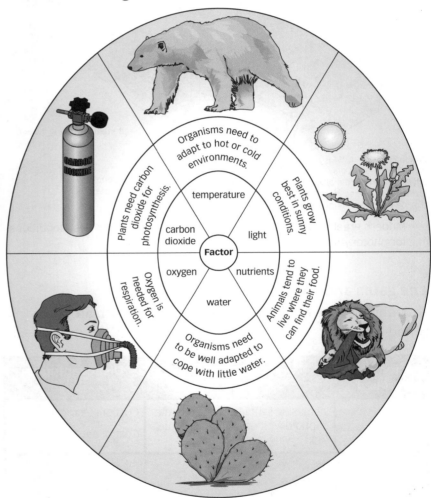

Sampling the environment

Biologists need to record data about the distribution of organisms. With this quantitative data they can begin to look for relationships concerning the cause of the distribution.

How can they record accurate unbiased data about an organism's distribution?

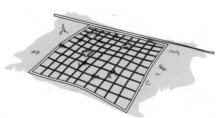

▲ A transect line is a long tape that is laid through an environment. Quadrats can be placed at regular intervals along the tape; this prevents bias. A quadrat is a square frame. They can be placed at regular intervals along a transect line to obtain unbiased data, or placed randomly in an area. The numbers and types of organism are recorded in the quadrat. This supplies the quantitative data.

Obtaining valid data

Being accurate	Use appropriate apparatus for the task, as this will generate accurate results. Each recording should be a sufficiently large sample.
Being reliable	Take repeat readings. Repeats make results more reliable.
Being fair	Always use the same equipment for each test. Make sure that recordings are not biased. To do this, use regular points along a transect, or random **sampling**.

If the data is collected in this way, it should be both valid and repeatable by other workers. Only then can any conclusions be accepted.

Analysing the data

Data collection of this type generates lots of numbers. Biologists often analyse the data to make sense of it. This may mean looking for a central value to illustrate the data, for example, the **mean** size of a limpet's shell.

There are three types of central values:
- **Mean** – this is the average value.
- **Median** – this is the middle value of the data when ranked.
- **Mode** – this is the most common value.

Making sense of the data

An example of this is where biologists have tried to explain the changing distribution of a species in the UK, like the ringed plover.

Collecting data	Analysing the data	Conclusions
They recorded the temperatures and numbers of plovers in the UK each month, for several years.	They averaged the temperatures for each month and averaged the numbers of plovers in the UK. They plotted a graph with the data over several years.	They looked for trends, patterns, or relationships in the data. For example, the increase in temperature has led to a change in the distribution of the species.

It is always important to make sure that the method used is valid, otherwise it can lead to incorrect conclusions being drawn.

Questions

1 How might you collect data about the distribution of organisms?

2 What three centralising values are used to analyse data?

3 What controls where an organism might live?

Working to Grade E

1   What are the raw materials needed for photosynthesis?

2   What are the **two** products of photosynthesis?

3   What is the source of energy for the reactions of photosynthesis?

4   Name the pigment that traps the energy of photosynthesis.

5   Which organ of the plant is the site of most of the photosynthesis in the plant?

6   Name **two** groups of organisms that carry out photosynthesis.

7   Below is a drawing of a section through the leaf.

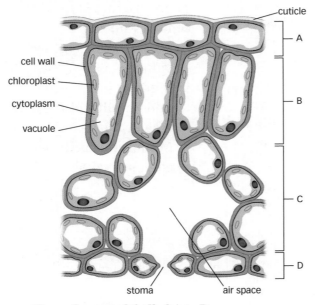

a   Name the parts labelled A to D.
b   Which tissue in the leaf is where most photosynthesis occurs?

8   What is a population?

9   Light is important to plants.
    a   How does light affect the distribution of plants?
    b   Explain why this is the case.

10 What is a quadrat used for?

11 What is the relationship between temperature and the distribution of polar bears?

12 What is a community?

Working to Grade C

13 Write out the word equation for photosynthesis.

14 Look back at question 7. Where does carbon dioxide enter the leaf?

15 What do we mean by the term 'limiting factor'?

16 Define the term 'rate of photosynthesis'.

17 Grass plants grow on a roadside verge.
    a   On a typical sunny day in June, what is likely to be the limiting factor for a plant?
    b   On a frosty day in November, is the same factor still likely to be the one limiting the plant?

18 Look at this drawing of a greenhouse.

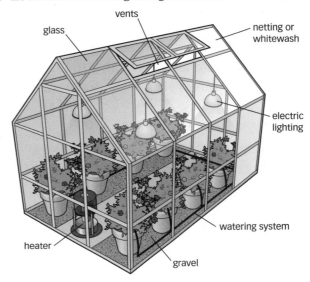

a   A paraffin heater is shown in the greenhouse.
    i   What **two** things are supplied to the greenhouse by the use of a paraffin heater?
    ii   Explain why these products of the paraffin heater should be important to increase plant growth.
b   Explain why it is important that the windows in the greenhouse can open.
c   Why is there gravel under the pots in the greenhouse?

19 Why is electric lighting often not considered commercially viable in greenhouses?

20 Why do organisms have to be very highly adapted to live in a desert?

21 A group of students decided to study the distribution of daisies in a school playing field. They worked in three groups; each placed a quadrat at regular intervals starting at the school building and working out into the open field. They calculated the average of their data. The results are below.

Distance from school (m)	Average number of daisies
5	5.2
10	5.8
15	8.9
20	20.1
25	25.3
30	28.3
35	27.0

  a Why did the students take three sets of readings for each distance?

  b How did the students ensure that there was no bias in the data they collected?

  c What trend can be seen in the data?

  d Suggest a reason for this trend.

22 Bison are herbivores.

  a What is the relationship between the distribution of the bison and the numbers of grassland plants?

  b If the bison overgraze the result is a decrease in the numbers of grass plants. What will happen to the distribution of the bison?

23 What is the difference between a mode and a median?

24 Scientific evidence must be valid and repeatable.

  a How does sample size affect validity?

  b Why is it important that experimental results are repeatable?

25 Look back at question 7.

  a Draw an arrow on the diagram to show the path taken by carbon dioxide to get to the photosynthesising cells.

  b Describe how the products of photosynthesis are removed from the leaf.

26 The products of photosynthesis can be converted into other molecules.

  a Proteins are one such molecule.

    i What is added to the products of photosynthesis to make a protein?

    ii Where does this additional substance come from?

  b What does the plant make to strengthen plant cell walls?

  c Why does the plant produce fats and oils?

  d What molecules make fruit sweet?

27 What substance does the plant use to carry out respiration?

28 Why is starch a more suitable storage compound than sugar?

29 The graph below shows how light intensity affects the rate of photosynthesis.

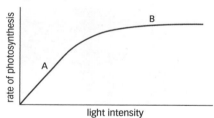

  a At which point on the graph, A or B, is light the limiting factor?

  b Explain why.

  c Why does the rate of photosynthesis not continue to increase?

30 Look at question 17, above.

  a When is the rate of photosynthesis going to be at its highest?

  b Explain your answer to question 17 part a.

31 Light is important to plants.

  a Explain how light is controlled in a greenhouse.

  b Explain why it is important for plant yield to control the light.

32 Explain how the use of a greenhouse to grow strawberries increases a farmer's profits.

33 Look at question 21, above. What other measurements would you need to take to prove your theory?

1 Use the words below to complete the diagram by filling in the empty boxes.

light	water	Sun	carbon dioxide	glucose	oxygen

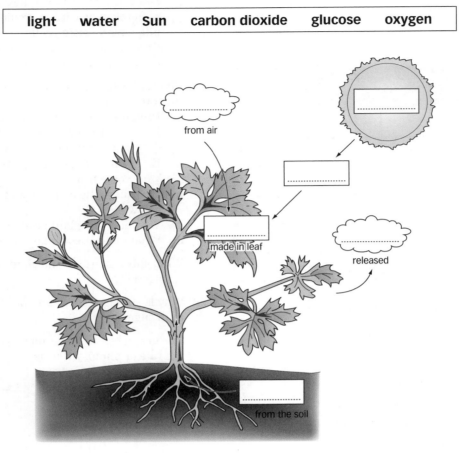

from air

made in leaf

released

from the soil

(6 marks)
(Total marks: 6)

2 A student investigated the effect of light intensity on the rate of photosynthesis. Below is a diagram of the apparatus that they used.

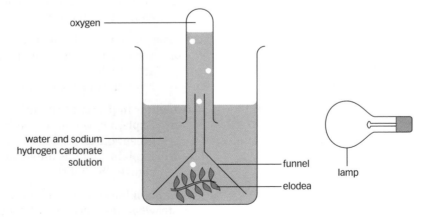

oxygen

water and sodium
hydrogen carbonate
solution

funnel

elodea

lamp

The student measured the light intensity, and the amount of oxygen produced as bubbles of oxygen produced per minute. The results are given below.

Light intensity (arbitrary units)	1	2	3	4	5	6	7
Number of bubbles of oxygen produced per minute	9	19		46	60	67	68

a Plot the data on the graph below, and join the points with a line.

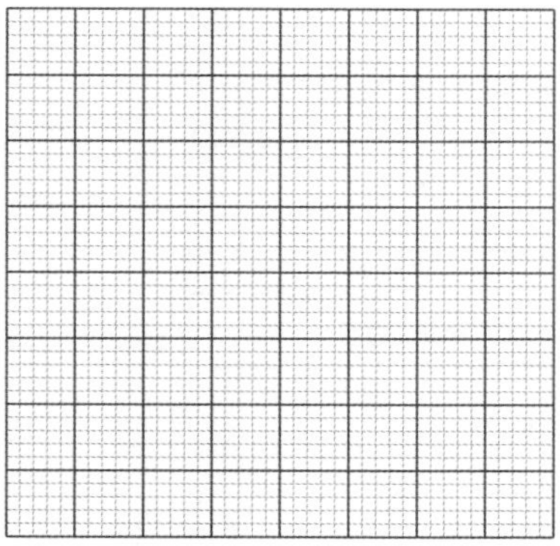

(3 marks)

b Estimate a value for the reading at a light intensity of 3.

....
(1 mark)

c Explain fully the relationship between light intensity and the rate of photosynthesis.

....

....

....

....

....
(3 marks)

d What is a limiting factor?

....

....

....
(2 marks)

e Name **two** other limiting factors of photosynthesis.

1 ....

2 ....
(2 marks)
(Total marks: 11)

3 A student set up an experiment to study the gases given off by leaves. They used a bicarbonate indicator that was sensitive to the levels of carbon dioxide.

The indicator shows the following colour range:

Purple	Very low carbon dioxide levels
Red	Medium carbon dioxide levels
Yellow	Very high carbon dioxide levels

Below are the results of the experiment.

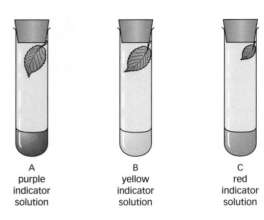

A	B	C
purple indicator solution	yellow indicator solution	red indicator solution

a Look at tubes A and B.

   i Which one of these tubes was kept in the light?

....

(1 mark)

   ii Explain your answer.

....

....

....

(3 marks)

b Why was it important that the two leaves in tubes A and B were the same size?

....

....

(1 mark)

c Why did the student avoid breathing heavily into the tubes when they set up the experiment?

....

....

....

(2 marks)

d Tube C was kept in the same conditions as tube A.

i Why was there a difference in the result?

....

....

(2 marks)

ii What might be the result in tube C if the leaf was killed by boiling before the experiment?

....

....

(1 mark)

(Total marks: 10)

4 A student wanted to conduct an experiment that would record the distribution of seaweed down a seashore.

a Describe a method the student could use to carry out this experiment. Identify any key equipment that might be needed.

....

....

....

....

....

....

....

....

....

(5 marks)

b How could the student ensure that the results would be:

i free from bias?

....

....

(1 mark)

ii reliable?

....

....

(1 mark)

(Total marks: 7)

Designing investigations to test a hypothesis

Scientists often start an investigation by suggesting a hypothesis. This might be a suggested relationship between two variables, in fact a prediction.

Whilst studying the ideas in this module, there are some good examples of investigations where you might be asked to produce a prediction that you could then test.

Look at the example that follows; this will show you how scientists design investigations to test their hypothesis. It will also highlight how scientists consider the validity of their data-collecting methods.

Distribution of Dog's mercury

Dog's mercury is a common woodland plant in Britain. It grows throughout the woodland, but seems to favour clearings rather more than ground under the tree's canopy.

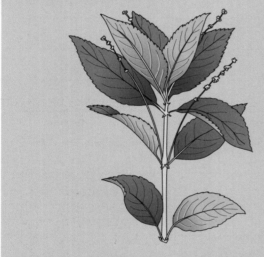

1   Propose a hypothesis for the distribution of Dog's mercury.

Skill – Producing a hypothesis

A hypothesis should have two parts:

A prediction – here it might be that Dog's mercury would grow better in higher light intensities.
- This identifies the two key variables: light and the growth of the plant.
- The prediction also gives a relationship between the two variables: more light = better plant growth.

A scientific explanation for the hypothesis – here it might be that where there is more light, there will be a higher rate of photosynthesis. This will make more food for the plant to grow.
- This explanation links a scientific process to the relationship given in the prediction.

2   In a woodland, your task is to design an investigation to test the hypothesis.
   a   What are the variables that you will measure?

Here you need to identify the variables:
- light intensity
- number of plants/percentage cover of Dog's mercury.

The two variables should come from your prediction.

   b   Given the following apparatus – a quadrat, a transect line, and a light meter – outline the method you would use to test the hypothesis.

Questions that ask you to design a method are looking for several clear and logical steps to be suggested. They should come from information on three key points:
- Do you know what each piece of apparatus is used for?
   - Place a transect line through the wood.
   - Place a quadrat at regular intervals along the transect line.
- Record quantitative data from the quadrats.
   - Take light readings in each quadrat.
   - Record the number of Dog's mercury plants or percentage cover of the plant in each quadrat.
- What will you keep the same?
   Include any two of the following:
   - size of the quadrat
   - distance between the quadrats
   - number of light readings in each quadrat
   - method for recording the plants in each quadrat.

   c   How would your method ensure that your results were valid in the following ways:
   - avoiding bias?
   - producing reliable, reproducible results?

To avoid bias you need to have a random system of taking your readings. This could be by placing quadrats randomly, or by placing them at regular pre-fixed points, such as at fixed distances from each other.

Reliable or reproducible results are produced by having sufficient repeats. (Remember 'For reliability do repeats'.) That will give you a large sample size in your results.

Answering a question where command words are important

In this country most tomatoes are grown commercially in greenhouses.

1 Name **three** conditions that can be controlled in greenhouses. *(1 mark)*

2 Describe how you would control one of these factors in a commercial greenhouse throughout the year. *(2 marks)*

3 Explain how this factor is important in the growth of the plant and yield of tomatoes. *(2 marks)*

QUESTION

G–E

1 Light, or when you burn a paraffin heater which helps plants grow.
2 Add more light.
3 It is important to put lights on in the greenhouse for the plants.

Examiner: Whilst light is the name of a factor the second part of the answer is not a name, but the description of a process that might be carried out. No third factor has been named. All three factors are needed for the mark.

In this answer the candidate has not obeyed the command word. They have not described *how* to control the condition. There is no thought of the times of the year. No marks awarded.

This candidate has not given an explanation. They might have picked up on the wrong word as a command. The command word here is *explain*. Command words always start the question. This answer is more of a description. No marks awarded.

D–C

1 Light, warmth, and moisture.
2 I would control the temperature by using a heater in the greenhouse for the winter, when it is cold. Burning the fossil fuel will add heat. This is good because it will also add carbon dioxide to help the plants to grow.
3 If it is warm, but not too hot, the plants will grow at its best for us, because of more photosynthesis.

Examiner: Clear, but warmth is not as precise as temperature, as temperature is the specific name of a factor that can be controlled. Three factors identified.

Here the candidate has described the use of a method to control the temperature during the winter, and so has picked up on the command word. Unfortunately they have not covered the entire year, and not discussed temperature control in the summer. There is also considerable irrelevant information in the answer, which deviates from the factor, and from a simple description. 1 mark.

A fair answer. The candidate has explained that the factor is used in photosynthesis, although some of their grammar is not the clearest. This would have gained one mark. The second mark has been lost because the student has not linked the photosynthesis to increased yield.

B–A*

1 Light, temperature, and moisture.
2 You could control the amount of light in the greenhouse by using nets under the glass in the summer when the sun is bright, and by using electric lights during the dark winter days.
3 Light is needed for photosynthesis in the plant. The more light the plant gets the faster the rate of photosynthesis. Then there will be more tomatoes.

Examiner: Clear and precise. Three factors identified.

This is a good answer. The candidate has clearly read and dissected the question. The command word asks for a description of *how* to control the factor. The student has talked of using nets and lights. The student has also picked up on the fact that the question asks for the whole year, and talked about the action taken in summer and winter. They gain 2 marks.

The candidate has given an explanation that links the factor to photosynthesis, and increased yield. 2 marks.

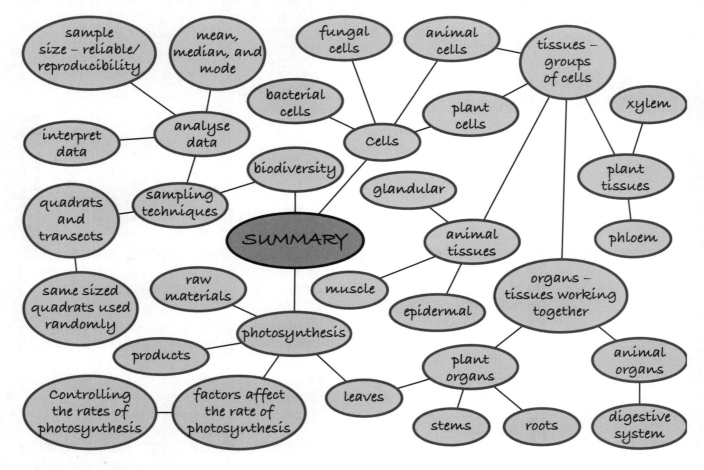

Revision checklist

- All living things are built out of cells.
- Cells contain parts that carry out different functions. Know the function of the nucleus, cytoplasm, cell membrane, cell wall, chloroplast, vacuole, ribosomes, and mitochondria.
- Know the difference between plant cells, animal cells, bacterial cells, and fungal cells.
- Cells become specialised for specific functions, for example, nerve cells can carry impulses.
- Diffusion is the process whereby molecules move into or out of cells. Diffusion is the random movement of molecules, and is affected by factors like distance, temperature, concentration gradient, and surface area.
- Similar cells work together in a tissue.
- Animal tissues include glandular, muscle, and covering. Plant tissues include xylem and phloem.
- Different tissues work together in an organ.
- Human organs include the stomach in which glandular, muscle, and covering tissue all work together. Plant organs include the leaves, stems, and roots.
- Different organs work together in an organ system to carry out a life process. The different organs might carry out different functions.
- An example of a human organ system is the digestive system. This system digests and absorbs food.

- Photosynthesis is the process by which plants make their own food. They use carbon dioxide from the air, light energy, and water from the soil, to make the waste product oxygen and the food glucose. Glucose can then be converted into other substances.
- The leaf is the organ in the plant involved with photosynthesis. It is highly adapted to carry out the function by being thin, having stomata for gaseous exchange, and having palisade cells with many chloroplasts.
- The rate of photosynthesis can be affected by light, temperature, and carbon dioxide levels. All of these factors can be controlled in a greenhouse.
- The distribution and biodiversity of organisms in the environment can be recorded by using sampling techniques. The numbers and location of the organisms can be measured using transect lines and regularly placed quadrats.
- Sampling techniques need to be valid by having a large sample size to make them repeatable and random or regularly placed sample locations to avoid bias. Often the data collected can be summarised by calculating a central value, such as a mean, mode, or median.

What is a protein?

Proteins are one of the major molecule groups that make up living things.

- They are built of **amino acids**.
- The amino acids are linked together in long chains.
- The chains are folded to give a specific shape.
- The shape is important for their function.
- The shape allows other molecules to fit into them.

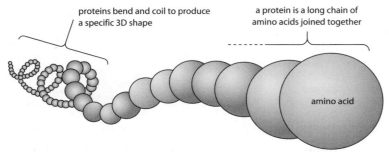

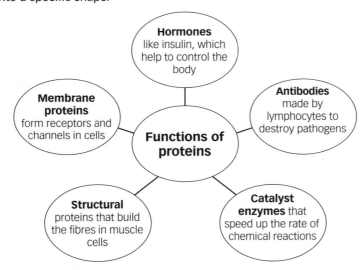

▲ Proteins are built from long chains of amino acids, which bend and coil into a specific shape.

proteins bend and coil to produce a specific 3D shape

a protein is a long chain of amino acids joined together

amino acid

Hormones like insulin, which help to control the body

Membrane proteins form receptors and channels in cells

Functions of proteins

Antibodies made by lymphocytes to destroy pathogens

Structural proteins that build the fibres in muscle cells

Catalyst enzymes that speed up the rate of chemical reactions

What are enzymes?

Enzymes are biological catalysts that speed up the rate of chemical reactions in the body.

- They are made of proteins.
- Without them the reactions of the body would be too slow for us to survive.
- The molecule that the enzyme works on is called the **substrate**.
- They can:
  - > break down large molecules into small ones, for example, in digestion
  - > build large molecules from small ones, for example, in photosynthesis.

Revision objectives

- ✔ know that proteins are made of long chains of amino acids
- ✔ understand a protein's shape allows it to carry out its function
- ✔ know that enzymes are proteins that catalyse chemical reactions
- ✔ understand that enzymes are specific, and work best at particular temperatures and pH

Student book references

2.14 Proteins

2.15 Enzymes

Specification key

✔ B2.5.2 a – b

Key words

protein, amino acid, hormone, antibody, catalyst, enzyme, substrate, specific, optimum, denatured

How enzymes work

The shape of the enzyme is vital for its function. The shape has an area into which substrate molecules can fit. This area is called the active site.

Substrate molecules fit into an enzyme's active site, for a reaction to occur. ▼

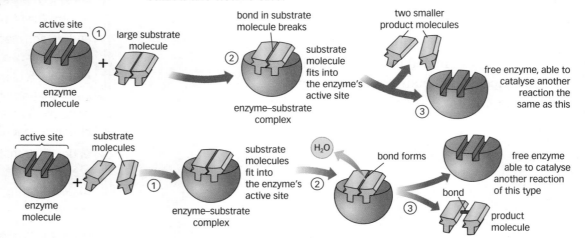

The key to the function of the enzyme is that the active-site shape is complementary to the substrate shape. This is not the same shape, but the two will fit together, like a key fits into a lock. No other substrate molecule will fit, which makes them **specific**.

What makes enzymes work best?

The way enzymes work is affected by temperature and pH.

Temperature

*   As the temperature increases, the rate of reaction increases. This is because the temperature causes the enzyme and substrate to move more and bump into each other more often.
*   This will not continue forever.
*   Eventually the rate reaches a peak called the **optimum** temperature.
*   Above the optimum, the increase in temperature starts to damage the shape of the enzyme. It cannot then work. The enzyme is said to be **denatured**.

pH

*   Each enzyme has an optimum pH. Here it works best.
*   Above or below this level it does not work so well. This is because the shape of the enzyme active site is damaged.
*   It is denatured.

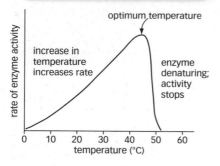

▲ Graph to show the effect of temperature on the rate of enzyme reactions.

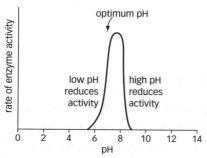

▲ Graph to show the effect of pH on the rate of enzyme reactions.

Questions

1	How are proteins constructed?
2	What **five** types of proteins function in the body?
3	**H** How do enzymes work?

Digestion

Digestion is the breakdown of large insoluble food molecules into small soluble molecules that can be absorbed into the blood. This provides the nutrients for us to survive.

Enzymes are important in digestion.
- They catalyse these breakdown reactions in the gut.
- They are produced by specialised cells in glandular tissue.
- They are released into the cavity of the gut.
- They then break down the food inside the gut.

The digestive process

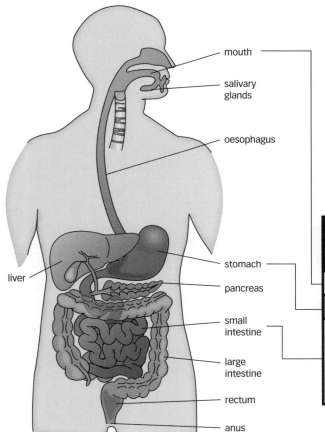

▲ Enzymes in the digestive process.

Gland where enzymes produced	Enzymes released	Reactions occurring
salivary gland	**amylase**	starch → sugars
wall of stomach	**protease**	proteins → amino acids
pancreas and small intestine	amylase	starch → sugars
	protease	proteins → amino acids
	lipase	lipids → fatty acids and glycerol (fats and oils)

Exam tip

Learn the three main digestive enzymes, where they are made, where they act, and what they do.

The importance of pH

The pH varies in the gut, and this affects which enzymes can work.

Acid pH in the stomach

The wall of the stomach produces **hydrochloric acid**. Only the stomach protease can work in this pH. The acid also helps by killing bacteria that enter the gut.

Neutralising the acid in the small intestine

The acidic food entering the small intestine from the stomach is neutralised by **bile**.

Bile is made in the liver → stored in the gall bladder → released into the small intestine → neutralises the acid.

This results in slightly alkaline conditions, which are best for the enzymes in the small intestine.

Questions

1   What is the process where food is broken down in the gut called?

2   What does amylase do?

3   What is the relationship between pH and enzyme action in the gut?

Working to Grade E

1 What are proteins made of?

2 Draw a diagram to show how the molecules that make proteins are arranged together.

3 Define an enzyme.

4 What is an enzyme made of?

5 Define digestion.

6 Bile is a digestive juice.

    a Where is bile made?

    b Where is bile stored?

Working to Grade C

7 Why is the structure of a protein important?

8 Complete the following table.

Type of protein	Function of protein
	Bonds to pathogens, destroying them.
Enzyme	
	Allows substances into cells through membranes.
	Controls the body's functions.

9 Give an example of structural proteins in the body.

10 Place the following statements into the correct order to explain the action of an enzyme.

    a The substrate molecules are brought together to form a bond.

    b The product leaves the active site.

    c Substrate joins the enzyme at its active site.

    d Once the bond is formed this is called the product.

    e The enzyme is once again available to perform another reaction.

11 What is the role of enzymes in:

    a digestion?

    b photosynthesis?

12 Temperature affects the rate of an enzyme-controlled reaction. What other factor will affect the rate of enzyme controlled reactions?

13 Why are enzymes vital for the survival of an organism?

14 Look at the table below and complete the blanks.

Region of enzyme action in the gut	Enzymes released	Reactions occurring
	amylase	starch → sugars
stomach		proteins → amino acids
		starch → sugars
	protease	
		lipids → fatty acids and glycerol (fats and oils)

15 How does hydrochloric acid help in the process of digestion?

16 How does the pancreas aid digestion?

Working to Grade A*

17 Explain what makes enzymes very specific.

18 Look at the graph below, which shows the change in the rate of an enzyme controlled reaction as the temperature changes.

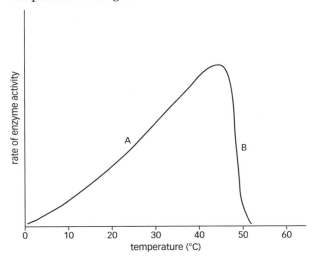

    a Describe how the rate of reaction changes as the temperature increases.

    b Explain why the rate is changing at point A.

    c Explain why the rate is changing at point B.

19 What is the role of bile in digestion?

Examination questions
Enzymes and digestion

1 Enzymes are involved in digestion. There are two types of protease involved in the digestion of proteins, but they work in different areas of the digestive system.

a What is the function of an enzyme?

Break dawn large malecules into smaller one and build large malecules fram small one

(1 mark)

b Look at the graph, which shows the rate of the two different enzymes at different pH values.

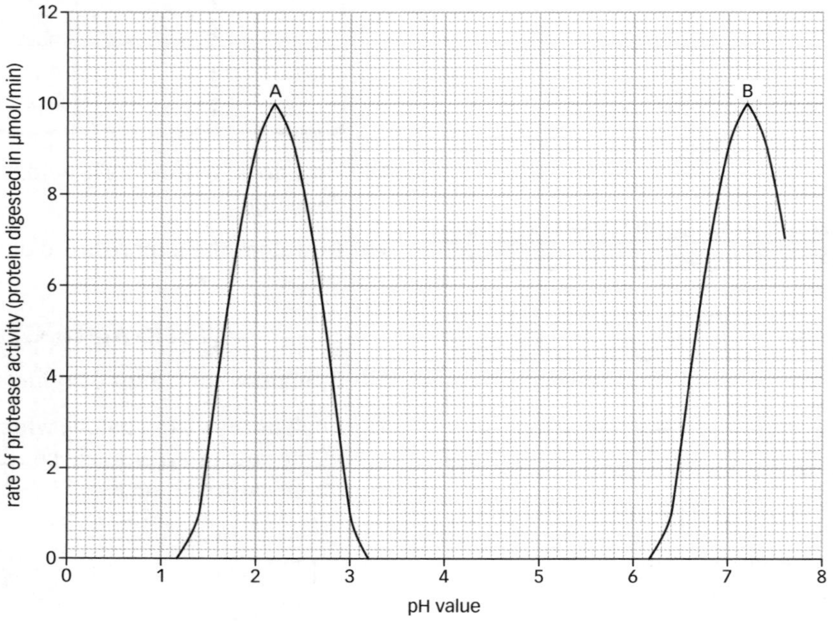

i In which region of the digestive system will enzyme A work?

....

(1 mark)

ii Explain your reasons for this answer.

....

....

....

(2 marks)

iii Complete the line for enzyme B to show how the rate changes as the pH increases.

(1 mark)

iv The optimum pH is the rate where the enzyme reaction is greatest. What is the optimum pH for enzyme A?

The aptimum ph far enzyme A is 10

(1 mark)

c Enzyme A does not work at the higher pH values of enzyme B. Explain in detail what has happened to the enzyme to prevent it working at this higher pH.

....

....

....

....

....

....

....

....

....

....

(4 marks)
(Total marks: 10)

Revision objectives

- ✔ know that enzymes from microorganisms are used in biological detergents
- ✔ understand the advantages and disadvantages of using enzymes in industry
- ✔ be aware of examples of the use of enzymes in industry

Student book references

2.17 Enzymes in the home – detergents

2.18 Enzymes in industry

Specification key

- ✔ B2.5.2 i – j

Obtaining commercial enzymes

Enzymes speed up the rate of chemical reactions. So they can be very useful in:

- the home – such as in washing powders
- industry – such as in food production.

We are able to obtain many of these enzymes from microorganisms grown in fermenters. The microbes make the enzymes and release them from their cells. We can then collect them for our use.

Biological washing powders

Detergents are cleaning agents. Detergents like washing powders are used to remove stains from clothes. Biological washing powders contain enzymes to help in the removal of stains. The advantage of these powders is that they remove stains that other powders leave behind.

stains on clothes are made mainly of proteins and fat

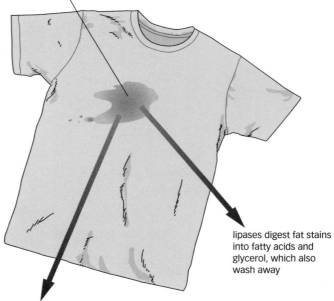

lipases digest fat stains into fatty acids and glycerol, which also wash away

proteases in the powder digest proteins into amino acids that wash away

▲ How enzymes work to remove stains.

No need to boil wash

Biological washing powders have a second great advantage. They can remove stains at lower temperatures than other washing powders. Most enzymes work best at 40 °C. So these powders will remove stains at these low temperatures and there is no need to boil wash. We can now wash more delicate fabrics, and save energy as well.

Enzymes in industry

Industrial processes are expensive because of the conditions they require. Enzymes are useful in industry because they allow reactions to occur:

- at lower temperatures
- at lower pressures.

This reduces the need for expensive, energy-demanding equipment.

Uses of enzymes

There are many examples of the use of enzymes. Different industries use different enzymes.

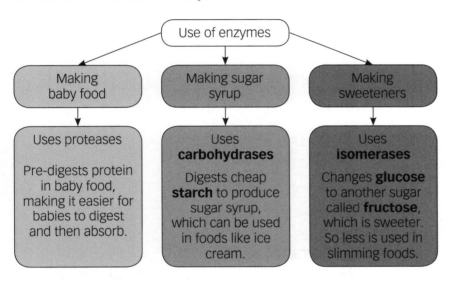

Disadvantages of using enzymes

Unfortunately it's not all good news! There are a few disadvantages with using enzymes in industry.

- The enzymes will be denatured at high temperatures. This means that their shape has changed and they can no longer work.
- They can be expensive to produce. This means that they need to save the industry a lot of money to be worth using at all.

Exam tip AQA

For each use of the enzymes there are advantages and disadvantages. These are all linked to the characteristics of enzymes learnt in B2 9.

Questions

1. What is the purpose of using enzymes in washing powders?

2. Name three enzymes used in industry.

3. **H** Explain why the temperature of industrial processes using enzymes needs to be maintained at 40 °C.

Revision objectives

- ✔ know that respiration is the release of energy, and occurs aerobically and anaerobically
- ✔ understand the process of aerobic respiration
- ✔ understand the uses of energy in plants and animals
- ✔ know the changes that occur in our bodies to allow energy release during exercise
- ✔ understand the process of anaerobic respiration

Student book references

2.19 Energy and life processes

2.20 Aerobic respiration

2.21 Anaerobic respiration

Specification key

✔ B2.6.1 ✔ B2.6.2

Key words

respiration, aerobic, anaerobic, mitochondria, exercise, glycogen, lactic acid, fatigue, oxygen debt

Respiration

Respiration is the process where cells release energy from molecules like sugar. It occurs continuously in both plants and animals.

It is controlled by enzymes. It can occur in two ways:

- **aerobic** – with oxygen
- **anaerobic** – without oxygen.

▲ Energy release from a sugar molecule.

Aerobic respiration

Aerobic respiration requires oxygen to release the energy from sugars like glucose.

- It is very efficient, releasing a lot of energy.
- The reactions of aerobic respiration occur mainly in tiny structures in the cell called **mitochondria**.

The equation for aerobic respiration is:

glucose + oxygen → carbon dioxide + water (+ energy)

How is the energy used by the organism?

Use	Explanation
Building larger molecules	Energy is used to build small molecules into larger ones.
	For example, amino acids are built into proteins in both plants and animals.
	Plants add nitrates to sugars and other nutrients to build amino acids.
Muscle contraction in animals	Energy is needed to cause the muscle proteins to contract. This will bring about movement.
Maintaining body temperature in mammals and birds	Energy is released as heat, which helps to keep the bodies of birds and mammals at a constant temperature. This keeps them active in colder surroundings.

Energy and exercise

Large amounts of energy are required in our bodies when we **exercise**. To release more energy a number of changes occur in our body.

The heart rate increases
- sending blood to the muscles quicker
- increasing the supply of oxygen and glucose to the muscles
- taking away the carbon dioxide

The breathing rate and depth increases
- taking more oxygen into the blood, to be sent to the muscles
- removing more carbon dioxide from the blood

A chemical called **glycogen**, which is stored in the muscle is broken down
- releasing more glucose into the blood
- sending glucose to the muscle for respiration

Anaerobic respiration

This type of respiration only occurs when there is not enough oxygen.

- It is less efficient at releasing energy than aerobic respiration.
- This is because it is an incomplete breakdown of glucose.
- The reactions occur in the cytoplasm of cells.
- This happens in muscles during intense or sprinting activities.
- The waste product is **lactic acid**.

$$\text{Glucose} \rightarrow \text{lactic acid (+ a little energy)}$$

Lactic acid is toxic and it builds up in muscles during long periods of vigorous exercise. When it reaches high levels it causes the muscles to become **fatigued**. This means they will no longer contract efficiently.

The blood flowing through the muscles will remove the lactic acid and take it to the liver where it will be broken down.

H   Oxygen debt

The only way to get rid of the toxic lactic acid is to use oxygen. So any build-up of lactic acid means the body needs oxygen; this is called **oxygen debt**.

The lactic acid produced in anaerobic respiration lowers the blood pH. This will cause an increase in breathing rate. This results in more oxygen being taken into the body, repaying the oxygen debt. The oxygen goes to the liver and the lactic acid is broken down into carbon dioxide and water.

Questions

1   Where does respiration occur?

2   What is the difference between aerobic and anaerobic respiration?

3   During exercise a number of changes happen in the body to help cells release more energy. Explain how **two** of these changes help in energy release.

Revision objectives

- ✔ know that mitosis is a type of cell division that produces identical body cells
- ✔ know the uses of mitosis
- ✔ know that meiosis produces gametes, which have half the chromosome number
- ✔ know the role of meiosis, and where it occurs

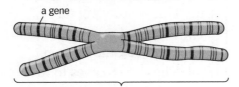

a gene

▲ Genes on a chromosome.

Exam tip   AQA

Questions often expect you to know the number of chromosomes in the new cells produced by either mitosis or meiosis. Remember in mitosis the number stays the same, in meiosis the number halves.

What are chromosomes?

Chromosomes are thread-like structures in the nucleus of every cell. Each chromosome contains many pieces of information called **genes**. Chromosomes are made of a chemical called DNA. In body cells chromosomes are found in pairs.

Mitosis

Body cells divide by **mitosis**. Before they can divide, each chromosome must make an exact copy of itself so that there will be one copy for each new cell. The chromosome then has an 'X' shape (as shown below). This process is called DNA replication.

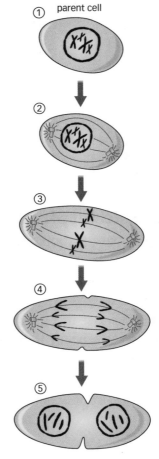

The 'X'-shaped chromosomes become visible.

Poles form at opposite ends of the cell, and the nucleus breaks up.

The 'X'-shaped chromosomes line up across the centre of the cell.

The chromosomes split, and one of each **identical** copy moves to opposite ends of the cell.

Two identical nuclei are formed. The cell divides once to form two genetically identical body cells.

Where does mitosis occur?

Mitosis occurs for:
- growth – new cells cause the body to get bigger
- repair – to replace old or damaged cells
- asexual reproduction – to produce new individuals, genetically identical (contain the same alleles) to each other and to their parent, called clones.

Gametes

Body cells always have chromosomes in pairs. During sexual reproduction **gametes** are made.

- Eggs are made in the ovaries of females.
- Sperm are made in the testes of males.

Gametes will only have one of the pair of chromosomes. To produce cells with only half the number of chromosomes, a second type of division is used, called **meiosis**.

H Meiosis

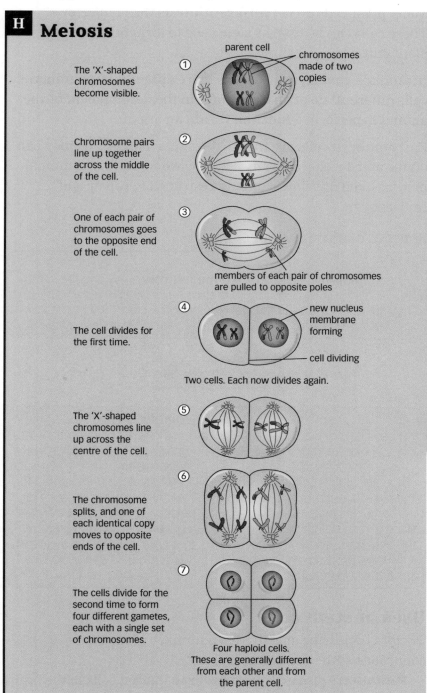

The 'X'-shaped chromosomes become visible.

① parent cell — chromosomes made of two copies

Chromosome pairs line up together across the middle of the cell.

②

One of each pair of chromosomes goes to the opposite end of the cell.

③ members of each pair of chromosomes are pulled to opposite poles

The cell divides for the first time.

④ new nucleus membrane forming — cell dividing

Two cells. Each now divides again.

The 'X'-shaped chromosomes line up across the centre of the cell.

⑤

The chromosome splits, and one of each identical copy moves to opposite ends of the cell.

⑥

The cells divide for the second time to form four different gametes, each with a single set of chromosomes.

⑦ Four haploid cells. These are generally different from each other and from the parent cell.

The role of meiosis

Meiosis is important because it halves the number of chromosomes. When two gametes **fertilise** each other, the normal adult number of chromosomes is regained.

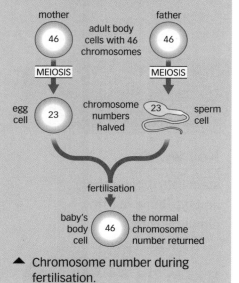

▲ Chromosome number during fertilisation.

Questions

1 What is a chromosome?

2 Where does mitosis occur?

3 **H** What is the difference between mitosis and meiosis?

This newly formed cell created through meiosis then repeatedly divides by mitosis to develop into a new individual.

What are stem cells?

Stem cells are undifferentiated cells. That means that they have not specialised as any one type of cell. These cells are useful to scientists because they can make them divide and differentiate into cell types that they need.

Where do we get stem cells from?

Many plant cells do not **differentiate** as the plant grows. These cells can be used as stem cells to form new roots on plant cuttings.

In animals, finding stem cells is more difficult. Most animal cells differentiate at an early stage in the development of the animal to perform a function within an organ.

Early **embryo** cells are the most useful stem cells as they can divide and develop into any cell type we want. As the animal matures, cell division is mainly restricted to repair and replacement.

Sources of animal stem cells

Adult stem cells
These are found in tissues like **bone marrow**. They are difficult to collect, and there are not many of them.
They are restricted as to the cell types they can form. There are fewer ethical concerns with this approach, as no life is destroyed.

Source of stem cells

Embryonic cells
These are the most useful as they will develop into any cell type.
There are ethical concerns with these cells as it means that the embryo has to be destroyed. Many people do not believe that the benefits outweigh the costs of such a process. They argue all life is valued.

Umbilical cord *blood cells*
These cells will differentiate into many cell types, and do not need the embryo to be killed. They may need to be stored for a long time. Whilst no life is destroyed, reducing the ethical concerns, some people worry that parents will have a child in order to be a donor of stem cells for others.

Uses of stem cells

Doctors hope to be able to treat a number of medical conditions with stem cells, such as:

• Parkinson's disease – by replacing damaged cells in the brain
• spinal injuries – replacing damaged nerves in the spine
• organ creation – for transplant
• diabetes – replacing damaged pancreas cells.

Working to Grade E

1 What is a detergent? ✗ lipase & protein

2 What **two** enzymes are used in washing powders?

3 The following statements refer to the advantages of using enzymes in industrial processes. For each statement, complete the sentence by using the word higher or lower as appropriate.
   a The reactions can be carried out at _____ temperatures.
   b The reactions can be carried out at _____ pressures.
   c The reactions will occur at a _____ rate.
   d The cost of the process will be _____ .

4 Define cellular respiration.

5 What is the source of energy in respiration?

6 Give **three** uses of the energy released by respiration in the body.

7 Energy is used during exercise. A number of changes occur in the body of an athlete to ensure that they can release energy. What happens to the athlete's heart rate during exercise?

8 There are two types of cell division: mitosis and meiosis. Which type of division is used:
   a when the body is growing?
   b to make gametes?
   c to repair damaged tissues?

9 Where are the gametes produced in males and females?

Working to Grade C

10 What substrates do the enzymes in washing powders work on?

11 What are the advantages of putting enzymes into washing powders?

12 Why are biological washing powders not effective in a boil wash?

13 Enzymes are used in industry.
   a For **each** of the following food products identify the:
      i substrate used
      ii enzyme
      iii product made.

   b Name a disadvantage of using enzymes in industry.

14 There are two types of respiration: aerobic and anaerobic. For **each** process, make a list of:
   a the reactants
   b the products.

15 There are two types of cell division: mitosis and meiosis. Which type of division is used during asexual reproduction?

16 Look at this drawing of an animal cell with four chromosomes.

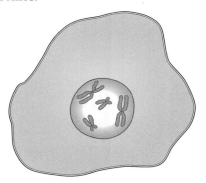

   a How many chromosomes will the daughter cells have if the cell divides by mitosis?
   b How many chromosomes will the daughter cells have if the cell divides by meiosis?
   c How many daughter cells will be produced if the cell divides by meiosis?

17 Name **two** sources of animal stem cells.

18 What is a stem cell?

19 Name a condition that stem cells may be used to treat in the future.

20 What is differentiation?

Working to Grade A*

21 Explain the reason why an enzyme is used in making the following products:
   a baby food.
   b slimming bar.

22 Explain how carbohydrases are used in the food industry.

23 Look at question 7, above. Explain why the athlete's heart rate changes during exercise.

24 What is oxygen debt?

25 What does the body do to relieve oxygen debt?

1   Respiration occurs in cells, when they release energy. Complete the following diagram to show what the cell needs, and the waste products it produces during respiration.

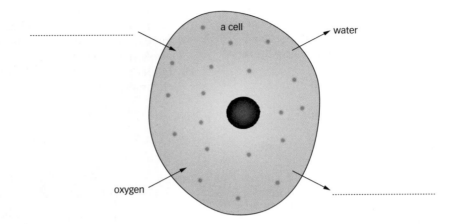

....   a cell   → water

oxygen

(2 marks)
(Total marks: 2)

2   Put the following parts in rank order of size, starting at the smallest.

| **Cell** | **Gene** | **Nucleus** | **Chromosome** |

......Gene...... → ......Chromosome...... → ......Nucleas...... → ......Cell......

(3 marks)
(Total marks: 3)

3   Below is a diagram of a type of cell division, showing the number of chromosomes in some of the cells.

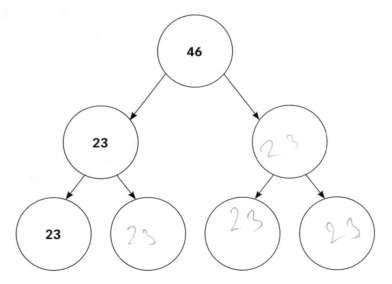

46

23   23

23   23   23   23

a   What type of division is this called?

......meiosis..

(1 mark)

b   Complete the diagram to show how many chromosomes are found in the other cells.   *(1 mark)*

c What type of cell is produced by this type of division?

gametes

(1 mark)

(Total marks: 3)

4 Stem cells are cells that can be grown in the laboratory, and are used by doctors to treat a number of conditions.

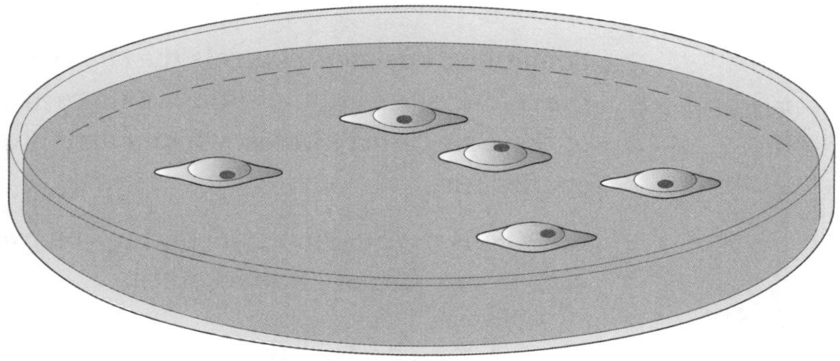

a Why are stem cells so useful to doctors?

they allow doctors to treat dead illnesses. They can develop & so treat illnesses

(1 mark)

b Name **one** source of stem cells.

bone marrow

(1 mark)

c Some people have expressed concern about the use of stem cells. Outline **one** argument that people can put forward against the use of stem cells.

(1 mark)

(Total marks: 3)

Revision objectives

- ✔ understand that there are different forms of genes called alleles
- ✔ be able to follow monohybrid inheritance
- ✔ be familiar with the work of Mendel on inheritance
- ✔ understand that the rediscovery of Mendel's work helped us to explain the processes of inheritance

Student book references

2.24 Inheritance – what Mendel did

2.25 The importance of Mendel's work

Specification key

✔ B2.7.2 a, c – e

Genes

Each individual is unique. They are a mix of characteristics from the father and mother. Characteristics are controlled by **genes**. Each gene is a section of DNA on a chromosome in the nucleus. Human body cells have 23 pairs of chromosomes. Eggs and sperm only have one of each pair.

- We inherit half our genes from our mother in the egg.
- We inherit half our genes from our father in the sperm.

Sometimes one gene controls the characteristic, for example, the colour of an animal's fur, but there are different versions of the gene called **alleles**, which might give the animal spotted fur.

Body cells have a pair of alleles for a characteristic, one on each of one pair of chromosomes.

Tracking inheritance

Inheritance is the passing of characteristics from one generation to the next. This means that genes are passed on. It is possible to follow the gene from one generation to the next. Following one characteristic like this is called monohybrid inheritance. Take this example of an animal's fur colouring:

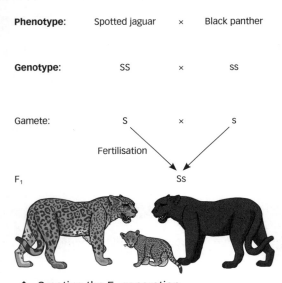

Phenotype:	Spotted jaguar	×	Black panther
Genotype:	SS	×	ss
Gamete:	S	×	s
		Fertilisation	
F₁			Ss

The phenotype is a description of the characteristic in words.

The genotype is a list of the genes present as a code, here S = spotted and s = black.

When the egg and sperm are made, only one of the alleles is present.

The two different alleles are brought together in the next generation, called the F₁. Here the animal is spotted. This is because the S allele is **dominant** and controls the development. The s allele is **recessive** and will only control development if the dominant allele is not there.

▲ Creating the F₁ generation.

H Two important words to know here are:

- **Homozygous** – here the genotype has identical alleles, for example, SS.
- **Heterozygous** – here the genotype has different alleles, for example, Ss.

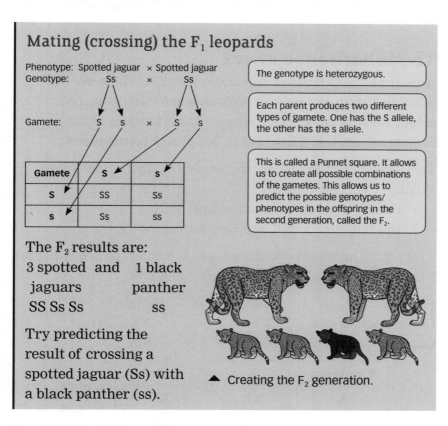

Mating (crossing) the F₁ leopards

Phenotype: Spotted jaguar × Spotted jaguar
Genotype: Ss   ×   Ss

Gamete: S   s   ×   S   s

Gamete	S	s
S	SS	Ss
s	Ss	ss

> The genotype is heterozygous.

> Each parent produces two different types of gamete. One has the S allele, the other has the s allele.

> This is called a Punnet square. It allows us to create all possible combinations of the gametes. This allows us to predict the possible genotypes/phenotypes in the offspring in the second generation, called the F₂.

The F₂ results are:

3 spotted and 1 black
  jaguars   panther
SS Ss Ss   ss

Try predicting the result of crossing a spotted jaguar (Ss) with a black panther (ss).

▲ Creating the F₂ generation.

Mendel – the father of inheritance

Gregor **Mendel** was born in 1822. He made a number of startling observations, which form the basis of our understanding of genetics. However, the importance of his work was not realised until after his death.

Mendel carried out many breeding experiments on pea plants, controlling the transfer of pollen from one plant to another. He was controlling the crossing of alleles, although he did not realise it. He worked at a time when scientists had not discovered chromosomes and certainly had not linked inheritance to them. As there was a lack of scientific knowledge at the time, the importance of Mendel's work was overlooked.

- He thought that inheritance of a characteristic was controlled by **factors** (we now call these genes).
- He worked out that the 'factors' must be in pairs in the adult cells.
- Only one of the factors would be in the gamete.
- The offspring would contain two factors, one from each parent.
- He was able to predict the outcome of crosses, just as we have shown above.
- The ratio of 3:1 in the F₂ is now called a Mendelian ratio.
- Finally his work was understood about 40 years later when other scientists could explain his results using genes.

Questions

1 What is inheritance?

2 Who was the pioneer of genetics?

3 **H** What genetic cross will always produce a 3:1 ratio in the offspring?

Revision objectives

- understand how sex is inherited
- know the structure and function of DNA
- know that DNA fingerprinting is a method for identifying people
- know about some genetic disorders that are inherited

Specification key

- B2.7.2 a – b, f – i
- B2.7.3

The inheritance of sex

Humans have 23 pairs of chromosomes; 22 of them are matching pairs and control body characteristics like hair colour and eye colour. One pair contains the genes that determine our sex. These are called **sex chromosomes**.

There are two different sex chromosomes, the larger X and smaller Y chromosomes.

- Females have two X chromosomes – XX
- Males have one X and one Y chromosome – XY

Every time humans reproduce there is a 50:50 chance of having a boy or a girl, as shown in the diagram below.

	Male	×	Female
Parents			
Chromosomes	XY	×	XX
Gametes	X Y	×	X X

Gametes	X	X	
X	XX	XX	½ girls
Y	XY	XY	½ boys

How does DNA work?

As we have learnt, chromosomes are made of a chemical called deoxyribonucleic acid (DNA). DNA is a spiral molecule, 'like a twisted ladder' called a double helix. Genes are just small sections of the DNA.

◀ Simplified diagram of a DNA molecule.

Each gene works by coding for the sequence of amino acids in a **protein**.

- The genetic code is contained in a sequence of **bases** in the DNA molecule.
- There are four different bases: A, T, C, and G.
- Each gene is made of hundreds of bases in a sequence.
- The cell reads the base sequence, three bases at a time.
- Each set of three bases is called a triplet.
- Each triplet codes for a specific amino acid.
- Each specific amino acid is joined to the next, and gradually builds a protein.
- The order of the amino acids is determined by the sequence of bases in the DNA.

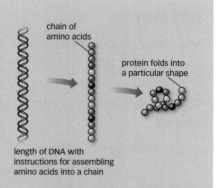

length of DNA with instructions for assembling amino acids into a chain

▲ Simplified diagram to show how the coded information in a gene determines the shape and the function of a protein.

DNA fingerprinting

Everyone's DNA is unique, apart from identical twins. This fact can be used to identify individuals. The technique of **DNA fingerprinting** was developed in the 1980s and has two major uses:

- to establish family connections like paternity
- to identify a criminal from evidence found at a crime scene.

The DNA sample collected is cut by enzymes, and separated to produce a series of bands on a gel. Scientists just need to match the bands to make an identification. Care is needed in the technique to avoid contaminating the sample.

▲ The banding pattern of a DNA fingerprint.

Inherited disorders

Sometimes a gene has a defect that results in a **genetic disorder**. There are about 5000 genetic disorders. Two examples are **polydactyly** and **cystic fibrosis**.

Polydactyly

This is a condition where additional digits develop on the hands or feet. It is caused by a dominant allele, and can then be passed on by one parent who has the disorder.

Cystic fibrosis

This is a disorder of the cell membranes. It results in a mucus build-up in the lungs. It can make sufferers more vulnerable to chest infections. Other organs like the pancreas can be affected, which might affect the digestive process.

It is caused by a recessive allele (c); the normal allele would be dominant (C). So three possible genotypes can exist:

- normal individual – CC
- carrier (no symptoms, but with the allele) – Cc
- sufferer – cc.

To be a sufferer the allele must be inherited from both parents.

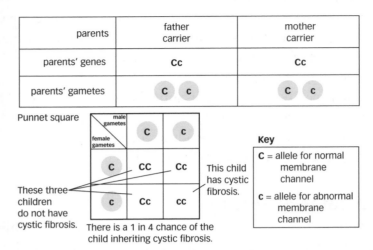

parents	father carrier	mother carrier
parents' genes	**Cc**	**Cc**
parents' gametes	C c	C c

Punnet square

These three children do not have cystic fibrosis.

This child has cystic fibrosis.

There is a 1 in 4 chance of the child inheriting cystic fibrosis.

Key

C = allele for normal membrane channel

c = allele for abnormal membrane channel

▲ A genetic diagram showing how cystic fibrosis is inherited.

Embryo screening

If both parents carry the cystic fibrosis allele, or any other genetic disorder, they may decide that they want children that are free of the disorder. To do this they would use in vitro fertilisation (IVF) to produce embryos. These embryos can be screened, and only those free of the disorder would be implanted.

Advantages of embryo screening	Disadvantages of embryo screening
The children won't have cystic fibrosis.	Embryos with the cystic fibrosis allele are discarded, so we are discarding life; is this ethical?
The cystic fibrosis allele won't be passed on to the next generation.	Some people with the genetic disorder feel that it is discrimination against them.
Saves money in the NHS because there will be fewer sufferers to treat.	

Questions

1. What determines whether we are male or female?

2. What causes cystic fibrosis?

3. **H** How does DNA work?

Evidence for previous life forms

Biologists believe that all organisms around today have developed from previous life forms. This theory is called **evolution**. Perhaps the best form of evidence for these earlier life forms is the **fossil** record.

Fossils are the preserved remains of living things from years ago. Fossils form in different ways.

Types of fossils

Remains of hard parts of animals like bones, which do not decay easily.

Preserved traces of organisms such as foot prints, burrows, and marks in the soil from roots.

Fossils

Remains of organisms that have not decayed, because a condition for decay is absent, for example, insects in amber.

Parts, most often hard parts, that are replaced by other materials such as minerals, for example, shells of ammonites.

Soft-bodied organisms do not fossilise easily, because the bodies rot. This means we have a poor record of these organisms. They do leave some imprint fossils, which are fine traces in rocks formed from silt. Many of these rocks will have been affected by geological activity, which destroys the fossils. This has resulted in little evidence about the earliest life forms, so scientists are unsure about exactly how life began on Earth.

▲ A fossil ammonite.

Fossils can be dated and placed into the correct time sequence. This fossil record shows us how much or how little organisms have gradually changed over time. They can even show us how one species might have changed into another, which is clearly seen in the evolution of the horse.

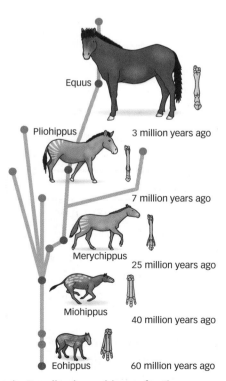

Equus

Pliohippus   3 million years ago

7 million years ago

Merychippus
25 million years ago

Miohippus   40 million years ago

Eohippus   60 million years ago

▲ Fossils give evidence for the evolution of the horse.

Revision objectives

- ✔ understand how fossils might be formed, and how useful fossils are to scientists
- ✔ know what extinction is and how it is caused
- ✔ understand the mechanism by which a new species develops

Student book references

2.30 Fossil evidence

2.31 Extinction

2.32 Forming new species

Specification key

✔ B2.8.1

Exam tip AQA

The fossil record is incomplete. You may be asked to evaluate fossil evidence in a question. Remember that uncertainty arises from the lack of reliable evidence because of an incomplete fossil record for most species.

Extinction

From the fossil record, biologists have found evidence of many examples of organisms that are not alive today. These organisms are extinct.

The causes of extinction

Cause	Effect
1 Changes to the environment over time	Some organisms are not well adapted to cope with the new conditions. For example, as global temperature increased, the woolly mammoths died out.
2 New predators	A more efficient predator hunts a population to extinction, for example, humans hunted dodos.
3 New diseases	Some diseases are so virulent that they destroy a population, for example, the elm tree is almost extinct because of Dutch elm disease.
4 New competitors	A new species may out-compete an existing one, for example, the grey squirrel out-competes the red squirrel, reducing the distribution of the red.
5 A single catastrophic event	Major global events such as asteroid impacts and volcanic eruptions dramatically change the environment. For example, an asteroid impact may have caused the extinction of the dinosaurs.
6 Speciation	As evolution naturally produces new varieties, some will be better adapted, and older versions may die out.

Speciation

Speciation is where one species evolves into two new species.

How new species form

Isolation
The population becomes divided into two groups by some kind of barrier like a new mountain range.

↓

H Genetic variation
Variations in characteristics develop in each separated population, caused by alleles.

↓

H Natural selection
In the two populations, different characteristics will be favoured in the different conditions. The individuals with these characteristics will survive. Their alleles are passed onto the next generation.

↓

H Speciation
Over time each group of the population becomes so different. Eventually they cannot interbreed, and so have formed new species.

Questions

1 What information do fossils give us?

2 Why are some scientists worried that global warming might lead to extinction?

3 **H** Explain the role of natural selection in speciation.

Working to Grade E

1 What is a gene?
2 Who was Mendel?
3 What organisms did Mendel work on?
4 Sex is determined by sex chromosomes.
   a What combination of sex chromosomes does a male have?
   b What combination of sex chromosomes does a female have?
   c Do the gametes of the male or the female contain different sex chromosomes?
5 What do the letters DNA stand for?
6 What is a genetic disorder?
7 Polydactyly is a genetic disorder caused by a dominant allele. What does polydactyly cause?
8 What is a fossil?
9 What is extinction?

Working to Grade C

10 Tongue rolling (T) in humans is dominant to non-tongue rolling (t).
   a In the following cross, show which genes would be in the gametes of the parents.

Parents:	Tongue roller	×	Non-tongue roller
Genes present:	TT	×	tt
Gametes:	◯ ◯		◯ ◯

   b Complete the genetic diagram below to show the outcome of a cross between the two parents.

Gametes		

11 In terms of genetics:
   a What does dominant mean?
   b What does recessive mean?
12 Chromosomes are found in the nucleus of human cells.
   a How many chromosomes are there in a human body cell?
   b How many chromosomes are there in a human sperm cell?
13 Sex is inherited.
   a Use the diagram below to show how sex is inherited.

Parents:	Male	×	Female
Chromosomes present:	_____	×	_____
Gametes:	◯ ◯		◯ ◯

Gametes		

   b What are the chances that a child born will be a boy?

14 Genes are located on chromosomes. The following diagram shows a pair of chromosomes. The gene for eye colour is marked on one chromosome. Where would it be on the other chromosome?

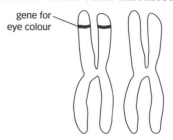

gene for eye colour

15 Describe the shape of the DNA molecule.
16 What does DNA code for?
17 Polydactyly is a genetic disorder caused by a dominant allele. Look at the family tree below for a family affected by polydactyly.

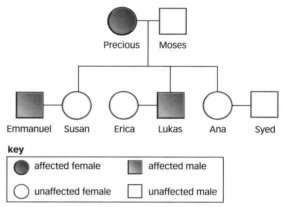

key

● affected female	▦ affected male
◯ unaffected female	☐ unaffected male

Moses does not have the symptoms. Does he have the polydactyly gene?
18 Are there any humans that might have identical DNA?
19 What uses are made of DNA fingerprinting?
20 Explain why soft-bodied organisms do not fossilise well.
21 Explain why a change in the environment might lead to the extinction of an organism.
22 What is speciation?
23 Why is isolation important for speciation?
24 Explain why scientists are uncertain about the origins of life.
25 How might humans have contributed to the extinction of the dodo?

Working to Grade A*

26 In terms of genetics:
   a what does genotype mean?
   b what does phenotype mean?
   c what does homozygous mean?
27 Brown fur colour (B) in mice is dominant to white (b). Construct a suitable genetic diagram to show the chances of two heterozygous mice producing a white mouse.

28 Mendel did some early work on genetics.
  a What were the findings of Mendel's work?
  b Why did Mendel struggle to get his work accepted?
  c Why do you think it was important that Mendel did not allow his plants to be naturally insect pollinated?

29 Explain why sexual reproduction produces variation in humans.

30 DNA contains four bases.
  a How many bases code for each amino acid?
  b What is the name given to the section of DNA that codes for one amino acid?

31 Explain how different genes code for different proteins.

32 Look at the family tree for a family affected by polydactyly in question 17, above.
  a What is the genotype of Precious and Moses?
  b Construct a genetic diagram to show the chances that a fourth child of Precious and Moses would have polydactyly.

33 Embryo screening is sometimes used by couples with a condition like cystic fibrosis in the family.
  a What is embryo screening?
  b Explain some of the concerns people have against embryo screening.

34 Explain why forensic detectives need to be so careful when collecting samples for DNA fingerprinting.

35 Look at this diagram of the evolution of the elephant family.

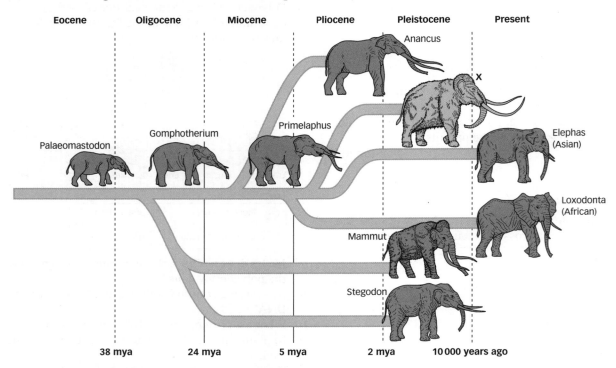

  a Many of these animals are now extinct. Biologists know of their existence by studying the fossil record. Which parts of these animals would fossilise best?
  b Explain why scientists cannot be absolutely certain about the evolutionary path of some of the elephants.
  c What traces might the elephants have left to allow biologists to explain how they walked?
  d The animal labelled X is a woolly mammoth. This became extinct many thousands of years ago. What is the most likely cause of its extinction?

36 A biologist called Alfred Russel Wallace studied animal species on either side of the Amazon River. He believed that two different species, one on each side of the river, evolved from the same common ancestor. Explain in detail how these two species might have developed.

Examination questions
Inheritance and evolution

1 Red flower colour (R) in geraniums is dominant to white flower colour (r). A plant grower crosses two geranium plants, one red and one white. All of the offspring are red.

    a   Use the following genetic diagram to show:

        i   the genes present in the parents   *(2 marks)*

        ii the genes in the gametes   *(2 marks)*

        iii the genes present in the offspring.   *(1 mark)*

Parents:	Red flowers	×	White flowers
Genes present:	………………..	×	……………………..
Genes in the gametes:	…………………		………………………

gametes		

        (Total marks: 5)

2 The giant panda is an animal that is in danger of becoming extinct.

    a   What do we mean by the word extinction?

………

………

        (1 mark)

b Read the following information about the life of the giant panda.

- The panda is a large land mammal.
- Its diet is composed of mainly bamboo shoots (about 99%).
- There are only about 2000 living in the wild.
- Each adult needs about 9–14 kg of bamboo shoots a day.
- Each bamboo species dies back regularly, and so a panda must have at least two species available as food.
- Pandas have a very low birth rate.
- The panda's habitat is being reduced by humans.

Use the information to suggest **three** reasons why the panda is in danger of extinction.

1 ....

....

2 ....

....

3 ....

....

(3 marks)
(Total marks: 4)

3 The Colorado River created the Grand Canyon in the United States of America. When this happened a species of squirrel evolved into two new species: the Abert and Kaibab tassel-eared squirrels on either side of the canyon.

Use your knowledge of speciation to explain how this might have occurred.

....

....

....

....

....

....

....

....

....

....

....

....

....

(5 marks)
(Total marks: 5)

Publish and share!

Mendel is regarded as the father of modern genetics. Today he is thought of as a major figure in the history of biology, but during his lifetime his work was rejected. How can this be? By looking at the way Mendel worked compared to modern scientists we can begin to understand why the importance of his discoveries was overlooked for many years.

Mendel's approach

Mendel set out to investigate the process of inheritance. His approach had four major steps.

The method

Mendel selected varieties of pea plants with clear characteristics. He was careful about controlling the process of pollination, to achieve only the cross he wanted. He avoided contamination by stray pollen from peas not involved in the experiment.

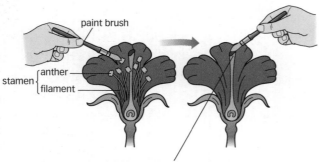

Pollen placed onto the stigma of another flower. This flower has had its stamens removed so it can only be pollinated by pollen from the other flower.

▲ Cross-pollinating flowers.

Mendel kept his experiments fair, by treating all plants in the same way after pollination, and growing all the produced seeds in the same way. He also carried out large numbers of experiments to collect sufficient data to make it reliable.

Data collection

He collected the seeds then grew them. The resulting crop showed different characteristics, and he meticulously counted the numbers of each type that grew.

Analysis of results

When he looked at the results he noticed that certain ratios always occurred, such as the 3:1 ratio. He was able to predict the result of genetic crosses.

Producing conclusions

Using his results he came up with a radical new idea for the process of inheritance. He argued that inheritance was not due to the blending of characteristics, which was the theory at the time. Briefly, he believed that:
- inheritance was controlled by factors
- factors were in pairs
- only one factor was in the gamete
- the offspring got one factor from each parent
- some factors were dominant.

What went wrong?

Failure to spread the word

Mendel tried to share his findings with the scientific community. However, he met with resistance. It was not in line with the ideas of the day, and many rejected the work because of that. Also there were gaps in his theory, as no one knew what these 'factors' were. This made scientists cautious. Mendel failed to get his ideas published widely for the scientific world to look at.

The consequences!

As the work was not published very few scientists knew anything of the work of Mendel. The problem with this was that no one could repeat the experiments to reproduce the results. In addition, other scientists were not aware of the idea of 'factors', so no one looked for them.

How would Mendel have been treated today?

Perhaps the biggest difference today is that scientists are able to publish, even if their results disagree with current views. The key issue is that the methods must be good and fair. Mendel's experimental technique seems sound.

Other scientists check their results. They will seek to reproduce similar results. Those experiments that do give the same results for other scientists will survive. If the experiment cannot be repeated then the ideas are dropped. Mendel's results can be easily repeated.

Today, if we have results that can be reproduced, but not fully explained, like Mendel's, we investigate them further. For example, scientists at the time did not know what factors were. Now scientists all around the world would read about new work, and where there were gaps in the knowledge they would try to explain them. Regarding Mendel's work, about 40 years after his work was written, other scientists found it, and were able to repeat it. Within a few years biologists had discovered the chromosome, and were able to rename factors as genes, which were located on chromosomes.

Answering questions where calculations and interpreting graphs are involved

QUESTION

The graph shows how a change in temperature affects the industrial digestion of starch into glucose.

1 The rate of activity is greatest at 40°C. What is this temperature known as? *(1 mark)*

2 Calculate the rate of production of glucose **per hour** at 30°C. *(2 marks)*

3 Explain why the rate of enzyme reactions starts to decrease after 40°C. *(3 marks)*

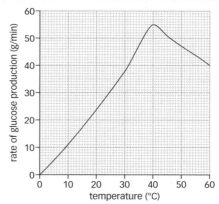

▲ Rate of glucose production from starch.

G–E

1 Body temperature.

2 40

3 The enzymes is being killed.

Examiner: No understanding of the functioning of enzymes here.

Unable to read accurately from the graph, and performed no calculation. Whilst the calculation is a higher-order skill, candidates must learn to read the graph accurately.

Incorrect technical vocabulary shows a lack of understanding; enzymes cannot be killed as they are not alive. No explanation is given, only a statement. 0 marks.

D–C

1 Maximum temperature.

2 38

3 At high temperatures the enzyme is denatured, so the rate drops.

Examiner: Some idea of the significance, but incorrectly recalled the technical keyword.

Correctly and accurately read the value from the graph, 1 mark, but they did not recognise that there was a second step in the calculation. The candidate must read the command words and hints in the question more carefully.

Short and succinct answer, but too brief to gain 3 marks. The candidate should note the number of marks available. They have an understanding of the technical ideas, but have not expressed them in enough points to pick up marks. This candidate scores only 1 mark.

B–A*

1 Optimum temperature.

2 38 × 60 = 2280 g/min.

3 At high temperatures the enzyme will lose its shape. This is called denaturation. When that happens the enzyme cannot fit to the substrate and so no reaction will occur.

Examiner: Good recall, correct use of technical vocabulary.

Correctly and accurately read the value from the graph, 1 mark, and performed the calculation for the second mark.

Well answered. Clear and logical. There is good use of technical vocabulary. There are four clear statements, 3 marks easily gained.

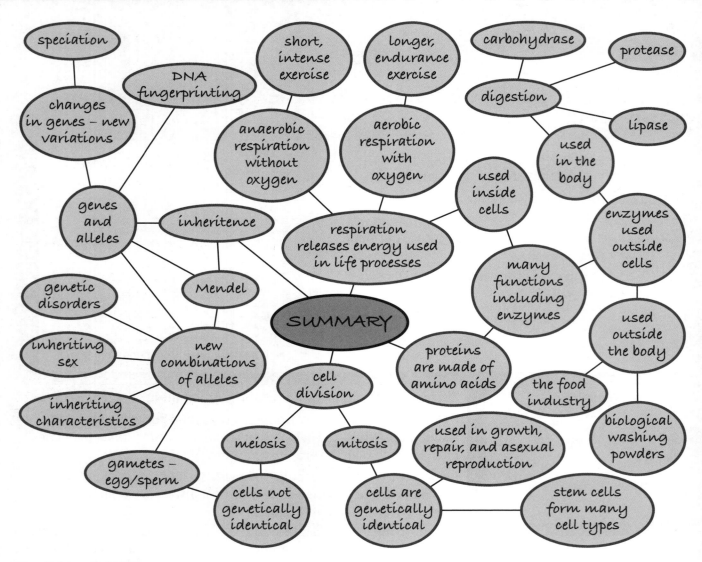

Revision checklist

- Proteins are important molecules in living things. They are made of amino acids and carry out many functions.
- Enzymes are proteins that catalyse specific reactions in living cells. Each enzyme works best at a specific temperature and pH.
- Enzymes are used in digestion to break down large molecules into smaller ones.
- There are many applications of enzymes: in the home in biological washing powders and in industry in the manufacture of baby food, glucose syrup, and diet foods.
- Respiration is the release of energy from foods.
- There are two types of respiration: aerobic, which uses oxygen and anaerobic, which does not use oxygen.
- Cells divide by mitosis during growth, repair, and also during asexual reproduction. The cells produced are genetically identical.
- Stem cells are undifferentiated cells, which can be triggered to differentiate into any cell type in the body. Stem cell technology could provide cures for many diseases, although there are ethical concerns.
- Meiosis is a second type of cell division, which is used in the production of gametes. It halves the chromosome number, and allows for genetic variation.

- Mendel discovered how characteristics are inherited. His ideas have formed the basis of modern genetics.
- There are two sex chromosomes (X and Y), which determine our sex. Males are XY and females are XX.
- Genetic diagrams can be used to track genes during reproduction, and allow us to predict the outcome of any genetic cross.
- Genes are made of DNA, and are found on chromosomes. Genes contain the code for the manufacture of proteins in the cell. There are different versions of genes called alleles.
- Some alleles of genes can lead to the production of abnormal proteins, resulting in genetic disorders. Embryos can be screened for the alleles responsible for these disorders.
- DNA fingerprinting is a technique that can be used to identify individuals because everyone (except identical twins) has unique DNA.
- Fossils are the remains of previous life forms. They allow us to see what these organisms looked like, and to be able to track how they have changed over time.
- Speciation is the process where new species develop over time, usually due to groups becoming isolated, and gradually changing.

Movement of molecules

Molecules can move into and out of cells by three methods. Dissolved substances move by:

- diffusion
- active transport.

Water moves by:

- **osmosis**.

Osmosis

Osmosis is a special kind of diffusion, which involves the movement of water only, into or out of cells.

During osmosis:

- water moves
- from an area of high water concentration (a dilute solution)
- to an area of low water concentration (a concentrated solution)
- through a **partially permeable membrane**
- until the concentrations even out.

Osmosis and cells

The movement of water into and out of cells by osmosis is important for both plant and animal cells because it keeps their water levels in balance.

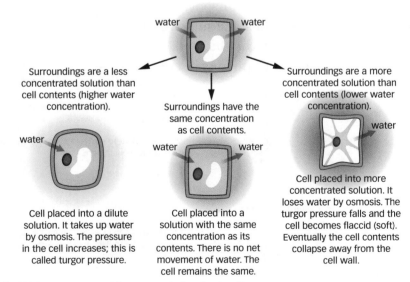

Surroundings are a less concentrated solution than cell contents (higher water concentration).

Surroundings have the same concentration as cell contents.

Surroundings are a more concentrated solution than cell contents (lower water concentration).

Cell placed into a dilute solution. It takes up water by osmosis. The pressure in the cell increases; this is called turgor pressure.

Cell placed into a solution with the same concentration as its contents. There is no net movement of water. The cell remains the same.

Cell placed into more concentrated solution. It loses water by osmosis. The turgor pressure falls and the cell becomes flaccid (soft). Eventually the cell contents collapse away from the cell wall.

▲ Water movement by osmosis in plant cells.

Revision objectives

- ✔ understand the process of osmosis
- ✔ know how exercise and sports drinks affect the hydration of the body
- ✔ know how the process of active transport occurs

Student book references

3.1 Osmosis
3.2 Sports drinks
3.3 Active transport

Specification key

- ✔ B3.1.1 a – g

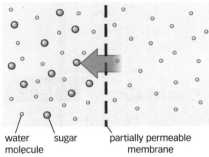

water molecule   sugar   partially permeable membrane

▲ Water is separated from a sugar solution by a partially permeable membrane. Sugar is too big to fit through the membrane pores but the water will pass through. So water molecules move into the sugar solution to dilute it.

Water to hydrate the body. Where dehydration is a problem, the more dilute drinks will hydrate the body quicker.

Caffeine to make us more alert.

Carbohydrate for energy. Moderate levels should be present to replace the glucose used during respiration. However, some drinks, called power drinks, have too much sugar and cause a sugar rush.

Ions are dissolved minerals to keep the muscles healthy. Ions are lost during sweating. Sports drinks should contain ions at the correct body levels, so the body will only absorb ions up to that level.

ENERGY

Exam tip   AQA

Learn the definition of osmosis. There are four key phrases in the definition, and many questions can be answered by using these four points.

Questions

1   What moves into a cell by osmosis?

2   What do sports drinks contain?

3   **H** What are the differences between osmosis and active transport?

The body and water

The human body contains a lot of water, which moves into the cells by osmosis. If the body's water level falls, the cells become **dehydrated**. It is important to replace lost water in the body by drinking.

A lot of water is lost from the body by sweating. This helps to cool us down. When we sweat we lose not only water but also ions from our bodies. The body's cells need water and ions to function correctly. Water and ions are also used in the body to:

*   lubricate joints
*   protect organs like the brain
*   carry substances around the body
*   help regulate body temperature.

Sports drinks

During exercise we sweat more to cool our body; this results in a greater loss of water and ions. In addition we also use a lot of energy, which we release from glucose.

Sports drinks are designed to solve these problems. However, the drinks vary and should be used for different situations.

Active transport

Some dissolved molecules or ions need to move into or out of cells from a low concentration to a high concentration, against a **concentration gradient**. This happens by a process called **active transport**. Active transport allows cells to absorb ions from very dilute solutions. Examples of active transport are:

*   the uptake of ions by root hairs
*   the movement of sodium ions out of nerve cells.

1. The ion attaches to the protein carrier.

protein carrier — sodium ion

cell membrane

There is a high concentration of sodium ions on the outside of the nerve, and a low concentration on the inside.

There are **protein carriers** in the nerve cell membrane. Sodium ions fit into these carriers.

2. The protein uses energy to change shape.

ATP

The proteins can change shape. This uses energy from a molecule called ATP, which is made in respiration.

3. The ion moves to the outside of the cell.

As the protein changes shape, it moves the sodium ion from the inside of the membrane to the outside.

The sodium ion falls off the protein carrier.

The protein immediately returns to its normal shape.

Working to Grade E

1 When we exercise will the following increase, decrease, or stay the same?
   a sweating
   b temperature
   c water intake
2 Name **three** ways molecules get into cells.
3 On a hot sunny day are you more likely to:
   a sweat **more/less**
   b become **hydrated/dehydrated**
   c become **more/less** thirsty
4 Does a dilute solution have a high or low water concentration?

Working to Grade C

5 Define osmosis.
6 What is a partially permeable membrane?
7 Look at the diagram below.
   a Indicate which direction the water will move.
   b Explain why the movement will occur in this direction.

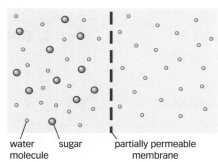

water   sugar   partially permeable
molecule   membrane

8 Give **three** reasons why it is important to keep the body hydrated.
9 Apart from water, what else does the body lose during sweating?
10 Sports drinks contain a number of different ingredients. Complete the following table to explain the function of each ingredient.

Ingredient	Function of ingredient
water	
sugar (carbohydrate)	
ions	
caffeine	

11 Suggest an advantage of drinking sports drinks rather than water during exercise.

Working to Grade A*

12 Look at the diagram in question 7. Explain why the sugar molecule does not move.
13 Describe what will happen to a potato chip after being placed for 24 hours in:
   a distilled water
   b a solution the same concentration as the cells
   c a very strong sugar solution.
14 Below is a table of the ingredients found in three different sports drinks. Look at the data and answer the questions that follow.

Drink	Sports Lite	Sports Active	Sports Power
water	250 cm³	250 cm³	250 cm³
glucose	2.5 g	8.7 g	22.0 g
ions	0.5 g	0.5 g	0.g
caffeine	0.0 g	0.0 g	3.0 g

   a Which of the three drinks would be better for hydration?
   b Explain why.
   c Explain why Sports Power would be better for a sprinter just before a race.
   d Explain why many people regard sports power drinks as unwise to take before sport.
15 Active transport is another form of transport.
   a Complete the paragraph below by using the most suitable words to fill in the blanks:
      Some dissolved molecules or ions move from a _____ concentration to a _____ concentration. The movement is _____ a _____ gradient. This process is called active transport.
   b Give **two** examples where active transport is used in living things.
   c For each example state clearly:
      i what is transported
      ii from where
      iii to where.
16 What **two** things are required for active transport to take place?
17 Explain why active transport will stop if a cell dies, or is given a respiratory poison.

1   A student set up an experiment where they placed a sugar solution inside a visking tubing bag and securely knotted the end. They then weighed the bag. The bag was placed into a beaker of distilled water for 30 minutes.

visking tubing bag

pure water

10% sugar solution

a   Describe what the result of the experiment would be after 30 minutes.

....

....

(1 mark)

b   Explain the reason for this result.

....

....

....

....

....

(3 marks)

(Total marks: 4)

2   Sports drinks are used by athletes.

a   Complete the following paragraph using some of the words supplied.

glucose   muscles   energy   fats   hydrate   dehydrate

Sports drinks contain carbohydrates, water, and ions. Carbohydrates such as .... are used in

respiration to provide .... Water is needed to .... the body. Mineral ions keep the

.... healthy.

(4 marks)

b   The table shows some of the contents of two different sports drinks. Indicate by using a tick (✓) to show what these sports drinks are best used for.

Drink contents	Hydration of the body	Energy supply for the body
Low sugar Dilute drink		
High sugar Concentrated drink		

(1 mark)

c   How is water lost from the body during exercise?

....

(1 mark)

(Total marks: 6)

Increasing in size

As an organism gets bigger, its **surface-area**-to-**volume** ratio gets smaller. This reduces the surface area over which molecules can be exchanged with the surroundings. Diffusion becomes inefficient.

surface area 6
volume 1

surface area 24
volume 8

surface area 96
volume 64

surface area : volume ratios
6 : 1
24 : 8 = 3 : 1
96 : 64 = 3 : 2 or 1.5 : 1

▲ The surface-area-to-volume ratio falls as a cell, or organism, gets bigger.

Exchange systems

Plants and animals have specialised, well-adapted **exchange surfaces**:

- the digestive system for nutrient uptake
- lungs or gills in animals for gas exchange
- the leaf in plants for gas exchange
- roots for water uptake.

Features of exchange systems

Efficient exchange surfaces have the following features:

- Large surface area – provides more surface for greater diffusion.
- Thin surface – provides a shorter distance for diffusion.
- Efficient blood supply in animals – maintains a concentration gradient.

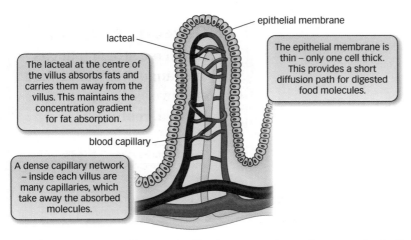

epithelial membrane

lacteal

The epithelial membrane is thin – only one cell thick. This provides a short diffusion path for digested food molecules.

The lacteal at the centre of the villus absorbs fats and carries them away from the villus. This maintains the concentration gradient for fat absorption.

blood capillary

A dense capillary network – inside each villus are many capillaries, which take away the absorbed molecules.

▲ Exchange across the villus. Each villus increases the surface area for both active transport and diffusion.

Questions

1 What is an exchange surface?

2 What molecules are exchanged across the villus?

3 **H** Describe the features of an efficient exchange surface.

Revision objectives

- ✔ know the location of the lungs and that they are the organs of gas exchange
- ✔ explain the process of gas exchange in the lungs
- ✔ describe the steps in the ventilation of the lungs

Student book references

3.5 Gaseous exchange in the lungs

Specification key

✔ B3.1.2

The lungs

The **lungs** are located in the chest or **thorax**, surrounded by the rib cage. The ribs protect the lungs and are also used in the process of breathing. The thorax is separated from the **abdomen** by a muscular sheet called the **diaphragm**. This encloses the lungs in the thorax.

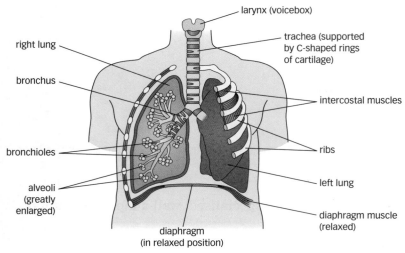

▲ This diagram shows a surface view of the left lung and a section through the right lung showing the airways and air sacs inside.

Into the lungs

Air gets into the lungs.

- Air moves in through the nose and mouth.
- It passes into a tube called the windpipe or trachea.
- The trachea branches into two tubes, each called a bronchus, one going to each lung.
- The bronchi divide into smaller and smaller tubes.
- Finally they end in the air sacs called alveoli.

Gaseous exchange

It is in the alveoli that **gas exchange** occurs. They are effective exchange surfaces. In the **alveolus**:

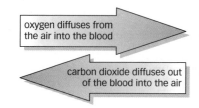

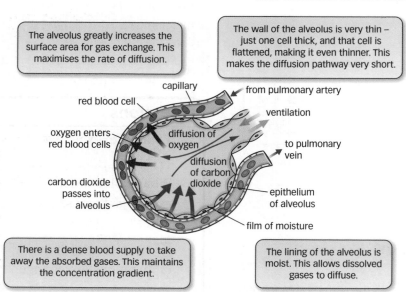

▲ How the alveoli and capillaries in the lungs aid gaseous exchange.

Ventilation

Ventilation is the movement of air into (inhaling) and out of (exhaling) the lungs.

Breathing in – inhaling

1 The intercostal muscles between the ribs contract, lifting the rib cage up and out, expanding the thorax.
2 The diaphragm muscle contracts, flattening the diaphragm. This also expands the thorax.
3 The volume inside the lungs increases, and the pressure decreases.
4 Air rushes into the lungs due to the low pressure.

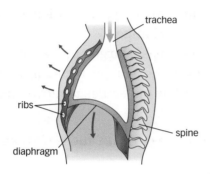

trachea

ribs

spine

diaphragm

Breathing out – exhaling

1 The intercostal muscles relax, and the ribs move down and in, reducing the volume of the thorax.
2 The diaphragm muscle relaxes, and arches up.
3 The volume inside the lungs decreases, which increases the pressure in the lungs.
4 The higher pressure forces air out of the lungs.

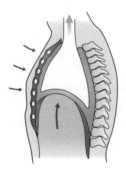

Some people experience difficulty in breathing. Asthmatics suffer from restriction of the bronchioles, which they treat with inhalers. These contain a drug to relax the bronchioles, which allows air to be inhaled and exhaled more freely. If ventilation stops, its action can be taken over by artificial ventilators to prevent damage to the organs of the body through lack of oxygen, and of course to preserve life.

Exam tip

You should be able to recognise and label the major organs of the respiratory system on a diagram.

Questions

1 What is the function of the diaphragm?

2 Which muscles move the ribs?

3 **H** What happens to the oxygen we breathe in?

Revision objectives

- ✓ know how plants exchange gases through their leaves, and the role of the stoma
- ✓ understand how water is absorbed and lost in transpiration
- ✓ explain the effects of environmental factors on the rate of transpiration

Plant exchange surfaces

Plants also need exchange surfaces. There are two major exchange surfaces in plants:

- leaves – where water vapour, carbon dioxide, and oxygen are exchanged with the air by diffusion
- roots – where water and minerals ions are absorbed.

Gas exchange in the leaf

The leaf is efficiently designed as an exchange surface:

- It has a flattened shape, giving a large surface area for **gas exchange**.
- It has many internal air spaces, which again increase the surface area for exchange.
- The lower surface has a large number of **stomata**. These are small pores in the leaf, protected by a pair of guard cells, which open to allow molecules to move into and out of the leaf.

Three gases move into or out of the leaf when the stomata are open.

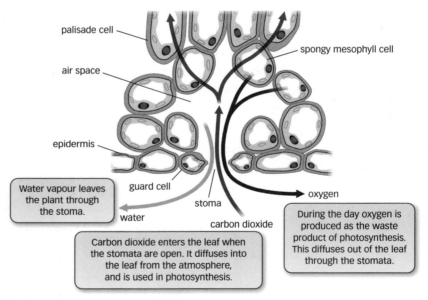

Water vapour leaves the plant through the stoma.

Carbon dioxide enters the leaf when the stomata are open. It diffuses into the leaf from the atmosphere, and is used in photosynthesis.

During the day oxygen is produced as the waste product of photosynthesis. This diffuses out of the leaf through the stomata.

▲ Gaseous exchange in the leaf.

The exchange of water

Water and mineral ions move into plants through the roots. The roots are adapted for exchange because they have root hairs, which increase the surface area for absorption.

water **evaporates** from the leaves – this process is called **transpiration**

water moves from the roots to the leaves – this is called the **transpiration stream**

water is absorbed by the root hairs

▲ The transpiration of water through a plant.

Controlling water loss

As well as allowing gaseous exchange, the stomata on a leaf control water loss.

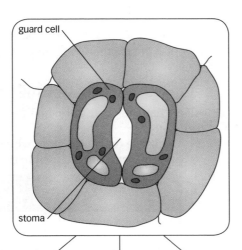

guard cell

stoma

Most stomata are on the lower surface where there is less heat from the Sun, to reduce water loss.

The guard cells can change shape to open and close the stoma. Stomata are usually open in the day to allow the gases in for photosynthesis, but closed at night to reduce water loss.

When the plants lose water faster than it is replaced from the roots, the stomata close to reduce water loss and to stop the plant wilting.

Factors that affect the rate of transpiration

There are four main factors that affect the **rate of transpiration**.

Factor	Effect on rate of transpiration
Increased light intensity	Increases – this is because stomata open in the light. As more stomata open, more water can evaporate, so transpiration increases. Eventually all the stomata open and the rate plateaus.
Increased temperature	Increases – the higher the temperature, the faster the water molecules in the air will move. This means they evaporate from the leaf quicker.
Increased air movement	Increases – when air moves over the leaf, it moves evaporated water away from it. The faster the air movement, the quicker the water is moved away, increasing the concentration gradient.
Increased humidity	Decreases – because if there are more water molecules in the air, the concentration gradient between the outside and inside of the leaf is reduced. It will take longer for water molecules to diffuse out of the leaf.

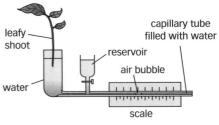

leafy shoot

water

capillary tube filled with water

reservoir

air bubble

scale

▲ The **potometer** is the piece of apparatus used to measure the rate of transpiration.

Questions

1 What is the role of the stomata?

2 List **three** factors that affect the rate of transpiration.

Working to Grade E

1 What happens to diffusion as the organism gets bigger?

2 Organisms have special exchange systems.
  a Why do organisms need such systems?
  b Give **three** examples of such systems, including at least **one** plant and **one** animal system.

3 Where are the lungs located?

4 The ribs surround the lungs. What **two** functions do the ribs have?

5 What **two** gases are exchanged across the lungs?

6 In which direction do the ribs move when we breathe out?

7 Where does gas exchange occur in the lungs?

8 Name the muscular sheet below the lungs, which separates them from the abdomen.

9 What is ventilation?

10 When we breathe in, the air passes through a sequence of structures. Put the following structures in the correct order.

| bronchus   alveolus   trachea   mouth |

11 Look at the diagram of the respiratory system below. Label parts A–E.

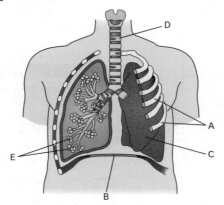

12 What are the **two** major exchange surfaces in the plant?

13 What is the transpiration stream?

14 Below is a drawing of a stoma shown in a section through the leaf.

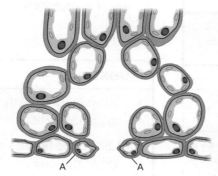

  a Label cells A on each side of the stoma.

---

  b Using arrows, show on the drawing the direction of movement through the stoma of:
    i oxygen
    ii carbon dioxide.

15 What piece of apparatus is used to measure the rate of transpiration in plants?

Working to Grade C

16 All exchange systems are modified to perform their functions.
  a Complete the table below, which shows how such systems are adapted to their functions.

Feature	How it improves diffusion
large surface area	
thin surface	
	to maintain a concentration gradient

  b Look at the diagram of the human villus below. Identify **three** features of the villus that are adaptations for efficient exchange.

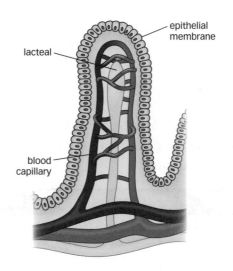

17 Explain why small organisms do not need an exchange system.

18 Give **three** adaptations of the lungs that make them an efficient exchange surface.

19 What causes the pressure to increase inside the lungs when we breathe out?

20 Complete the following table to say whether a gas has increased, decreased or stayed the same, in the air we breathe out, compared to the air we breathe in.

Gas	Change
oxygen	
carbon dioxide	
water vapour	
nitrogen	

21 List **three** factors that affect the rate of transpiration.

22 Stomata are the pores through which gas exchange occurs in the leaf.
  a On which surface of the leaf are stomata mainly found?
  b Explain why this is an advantage.

23 Look back at question 14.
  a Using an arrow, show on the drawing the direction of movement through the stoma of water.
  b When does the stoma mainly open?

24 Explain why it is an advantage for most leaves to have a large surface area.

25 Describe **one** way a root is adapted to be efficient at diffusion.

26 What happens to a plant that is unable to replace water as quickly as it is lost?

Working to Grade A*

27 Think about the features of the villus you identified in question 16b. Explain how these features make diffusion more efficient.

28 Below is a drawing of two cube-shaped cells of different sizes. The lengths of the sides of the cell are shown.

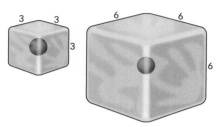

  a Complete the table below by calculating:
    i the surface area of the cells
    ii the volume of the cells
    iii the surface-area-to-volume ratio of the cells.

Cell dimensions	Surface area	Volume	Surface-area-to-volume ratio
3			
6			

  b What do you notice about the size of the ratio as the cell gets larger?
  c Explain why it is a problem to be big.

29 Explain how we inhale. In doing this your explanation should include the role of the following structures in the process:
  • ribs
  • diaphragm.
You should also describe what happens inside the lung to:
  • the volume
  • the pressure.

30 Most leaves have a large surface area.
Plants like Marram grass, which live on dry sand dunes, have reduced the surface area of their leaves. Explain why.

31 Describe the relationship between the rate of transpiration and the following factors:
  a light intensity
  b air movements
  c humidity

Examination questions
Exchange surfaces in the lungs and in plants

1 Two students decide to investigate breathing rates.

One student acts as the subject and the other records the number of breaths taken in 20 minutes by the other as 320.

a Calculate the breathing rate per minute. Show your working in the space below.

....

.... breaths per minute.
(2 marks)

b Predict what might happen to the breathing rate of the student if they begin to exercise.

....
(1 mark)

c Gas exchange takes place in the alveolus.

i What gas does the student remove from the air?

....
(1 mark)

ii Describe how the alveolus is adapted to be an efficient exchange surface.

....

....

....

....

....
(3 marks)
(Total marks: 7)

2 A plant biologist designed an experiment to investigate the effect of environmental conditions on the rate of water loss in plants.

The biologist recorded the rate of water uptake by the plants in different conditions of light and wind. The graph shows the results.

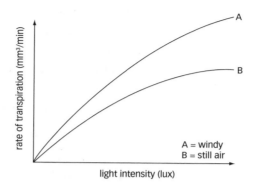

a Describe the relationship between the light intensity and the rate of water loss.

....

....

....

(1 mark)

b Water is lost through tiny pores called stomata on the leaf.

i What do you think happens to the stomata in the dark?

....

....

(1 mark)

ii What evidence allows you to think this?

....

....

....

(1 mark)

c Gardeners need to water their plants. Use the graph to explain in which weather conditions gardeners will need to water their plants most.

....

....

....

....

(2 marks)

(Total marks: 5)

The need for a circulatory system

As animals get larger they need a circulatory system. This is because diffusion becomes too inefficient to move molecules like:

- oxygen – from the surface, deep into the cells of the body
- waste carbon dioxide – from the cells, to the outside of the body
- foods – from the small intestine, to the cells of the body.

Circulatory systems transport these substances around the body.

Parts of a circulatory system

Human circulatory systems have three component parts:

- Blood – a fluid to carry the molecules.
- The **heart** – a pump to move the blood around the body.
- Vessels – tubes to contain the blood.

The human circulatory system

Humans have a double circulatory system. This means that the blood passes through the heart twice as it makes its way around the body. The heart pumps deoxygenated blood to the lungs in the first circuit, and oxygenated blood to the body in a second circuit.

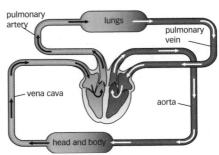

▲ The human double circulatory system.

In a complete **circulation** blood passes:

- from the heart to the lungs to remove carbon dioxide and collect oxygen
- back to the heart
- to the body organs and tissues
- back to the heart before going to the lungs again.

The heart

The heart is an organ that pumps blood around the body. Typically the heart beats 60–80 times a minute. Much of the wall of the heart is made of muscle tissue. The heart is divided into four chambers (left and right **atria** and left and right **ventricles**). The atria have thin walls as they only pump blood to the ventricles. The ventricles have thick walls as they pump blood all around the body.

▼ Section through the human heart showing circulation of the blood.

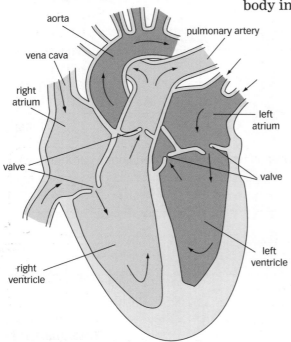

Circulation through the heart

1   Deoxygenated blood arrives from the body through the **vena cava** to the right atrium.
2   The right atrium contracts and forces the blood into the right ventricle.
3   The right ventricle contracts and forces the blood up and out of the heart through the **pulmonary artery**.
4   There is a **valve** between the ventricles and atria, which is forced shut when the ventricles contract, preventing backflow of blood, so the blood flows in the right direction.
5   A second valve prevents blood from the artery draining back into the heart.
6   Blood goes to the lungs and picks up oxygen and loses carbon dioxide.
7   Oxygenated blood returns to the left atrium of the heart through the **pulmonary vein**.
8   The atrium contracts and forces blood into the left ventricle.
9   The left ventricle contracts and forces the blood out of the heart through the **aorta**, to the body. The left ventricle has a thicker wall to pump the blood all around the body.
10 The two valves again prevent the backflow of blood in the heart.

These two processes happen at the same time: the atria contract together, then the ventricles contract together, so the process of blood moving round your body and through your lungs is a continuous flow.

Artificial hearts and heart valves

Common circulatory conditions include:
- Heart failure – where the heart muscle is damaged, and struggles to contract.
- Valve damage – where backflow of blood is not prevented.

Developments in biomedical and technological research enable us to help treat these conditions.

Artificial hearts have been developed. These can be used as a short-term solution to heart failure whilst waiting for a transplant. These keep the patient alive and are not rejected by the body. However, they often have wires that protrude through the skin. It is hoped that artificial hearts will improve.

Artificial valves are plastic and metal valves that can replace damaged valves. These are long lasting and highly successful.

Exam tip

Make sure you can identify and label the four chambers of the heart and the names of the blood vessels associated with them. Learn the circulation of blood through the heart as a sequence. You could learn them as series of numbers, or colour each statement with a sequence of colours.

Questions

1   Explain why we need a circulatory system.
2   What is the use of an artificial heart?
3   **H** How does the heart pump blood around the body?

The blood vessels

The blood vessels are the tubes through which the blood flows. There are three types of blood vessel.

Arteries take blood away from the heart, **capillaries** take blood through the organs, and **veins** return blood to the heart. Their structure is related to their functions.

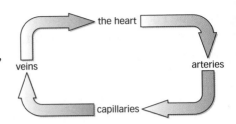

Arteries

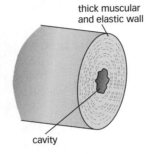

thick muscular and elastic wall

cavity

- walls are thick because the blood is under high pressure
- large amounts of **muscle** allow the wall to withstand and maintain the pressure
- large amounts of **elastic fibres** allow the artery to stretch and recoil as blood surges through
- narrow cavity or lumen

Capillaries

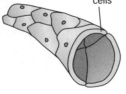

wall made of single layer of cells

- walls are very thin, only one cell thick, so diffusion is quick
- large number of capillaries gives a large surface area for diffusion
- molecules needed by the cells (such as oxygen and glucose) pass out of the blood
- molecules produced by the cells (carbon dioxide and wastes) pass into the blood
- blood pressure has been lost, and the blood flows slowly by the time the blood reaches the capillaries
- very narrow, just wide enough to allow one red blood cell through

Veins

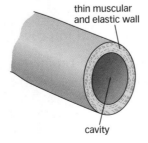

thin muscular and elastic wall

cavity

- thinner walls than arteries, because the blood pressure is lower
- little muscle or elastic fibres as there is no high pressure to withstand
- valves to prevent backflow of blood
- large lumen

Treating narrowed arteries

A common circulatory problem is narrowing of the arteries. This is usually due to the build-up of fat in the wall of the artery. If this occurs in the arteries supplying blood to the heart muscle (the coronary arteries), it could result in a heart attack. To treat this condition a wire mesh is inserted into the narrow region of the artery and expanded. This **stent** opens the artery.

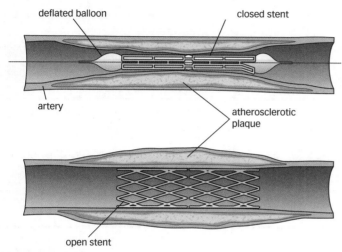

▲ A coronary stent.

Blood

Blood is a tissue, because it is made of similar cells working together. It is a fluid that flows through the blood vessels, pumped by the heart. It has three main functions:

- Transport – carries substances and cells around the body.
- Protection – from infection and blood loss.
- Regulation – helps to maintain the body temperature and pH.

Composition of the blood

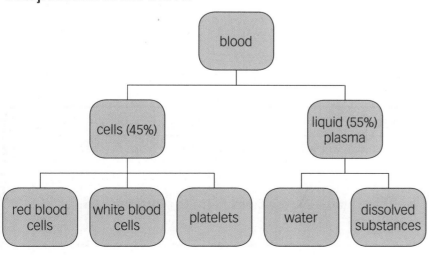

Exam tip AQA

Remember the functions of the blood vessels. Arteries always take blood away from the heart (**A** for **a**way). Ve**in**s take blood **in**to the heart.

Key words

artery, capillary, vein, muscle, elastic fibre, stent, plasma, haemoglobin

A closer look at the blood

Component	Function
Plasma	Transports dissolved substances such as: • carbon dioxide from the cells to the lungs • soluble products of digested foods from the small intestine to the rest of the body • urea from the liver to the kidneys.
Red blood cell	• Contains the red pigment called **haemoglobin**. This combines with oxygen in the lungs to form oxyhaemoglobin. • Red blood cells transport the oxygen around the body; oxyhaemoglobin then breaks up to release the oxygen in other organs. • There is no nucleus, which provides more room for haemoglobin. • Made in the bone marrow, and destroyed in the liver.
White blood cell nucleus	• There are several types. They all contain a nucleus. They all form part of the immune system, working to fight infection. • Some engulf and digest microorganisms, others make antibodies to destroy microorganisms.
Platelets	• Small fragments of cells with no nucleus. They help form blood clots at the site of wounds, to prevent blood loss and infection.

Artificial blood products

It is important to maintain our blood volume. When people have serious injuries, such as during war or major trauma, it can result in major blood loss. It is important to replace lost volume. When real blood is not available, artificial blood can be used. This will not be rejected by the body and will maintain blood pressure. Some types also contain chemicals to help transport oxygen.

Questions

1   Does an artery take blood to or away from the heart?

2   Why does an artery have a thicker wall than a vein?

3   H Explain how the blood helps to protect us against disease.

Plant transport systems

Plants also have transport systems. The separate systems involve two types of tissue called:

- xylem
- phloem.

They are located in a vascular bundle.

Summary of transport in plants

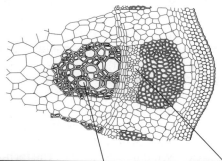

Tissue name	Xylem	Phloem
What does the system transport?	water and mineral ions	dissolved sugars
From where?	roots	leaves (source)
To where?	leaves	rest of plant (sink), including growing regions and storage organs
Cell structure	dead cells stacked on top of each other to form tubes	living cells stacked on top of each other to form tubes
Name of process	transpiration stream	translocation

A closer look at the transpiration stream

The diagram shows the detail of how water and minerals are transported from the root to the leaves.

upper skin of leaf

leaf vein

leaf

xylem vessels in the stem

stoma   guard cell

Water moves into the leaves. It evaporates from leaf cells and escapes through stomata as water vapour.

water and minerals

The root hair takes in water and dissolved minerals from the soil

Water and minerals move from cell to cell through the root until they reach xylem vessels

Water and minerals move up through the xylem vessels to the stem and the leaves

▲ The process of transpiration.

Revision objectives

- ✔ know the role of xylem in the transport of water and ions in plants
- ✔ know the role of phloem in the transport of sugars in plants

Student book references

3.14 Transport in plants

Specification key

✔ B3.2.3

Key words

xylem, phloem, transpiration stream, translocation

Exam tip AQA

This topic is about two tissue types with two separate functions: learn the table of comparisons above. Look back at transpiration in plants, and link the ideas to this spread.

Questions

1. How does the plant transport water?

2. How does the plant transport sugars?

3. **H** Where is sugar transported from and to?

1   What are the **three** major parts of the human circulatory system?
2   What gas does the blood pick up at the lungs?
3   Below is a diagram of the human heart.

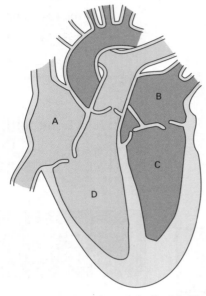

   a   Name the four chambers labelled A–D.
   b   Label the aorta.
   c   Label a valve.
4   Which type of blood vessel takes blood away from the heart?
5   Which type of blood vessel takes blood to the heart?
6   Name **two** substances transported in the plasma.
7   The blood contains a number of cells.
   a   What is the function of the red blood cell?
   b   What is the name of the pigment that helps the red blood cell carry out its function?
8   What is transported in the xylem?
9   What is transported in the phloem?

10   Why is the human circulatory system described as a double circulation?
11   What is the name of the blood vessels which take blood:
   a   to the lungs?
   b   to the heart from the body?
12   Look at the diagram in question 3. Use arrows to show the circulation of deoxygenated blood through the heart.
13   What is the function of the valves in the heart?
14   The heart chambers are surrounded by walls of muscle.
   a   Why are the walls of the ventricles thicker than the walls of the atria?
   b   Why is the wall of the left ventricle thicker than the right ventricle?

15   How would a surgeon treat a heart with a damaged valve?
16   Artificial hearts have been developed to treat heart failure.
   a   Give an advantage of using an artificial heart.
   b   Give a disadvantage of using an artificial heart.
17   The blood contains a number of cells.
   a   What type of cell produces antibodies?
   b   What is the function of a platelet?
   c   Why do red cells have no nucleus?
18   What are the **three** functions of the blood?
19   Artificial blood has now been developed. Give **one** situation where artificial blood might be used.
20   Stents are used to treat some circulatory problems. What is a stent?
21   Name a difference between xylem and phloem tissues.
22   What is translocation?
23   In translocation biologists often talk about substances moving from a source to a sink.
   a   What do we mean by the term 'source'?
   b   Give an example of a source.
   c   Give an example of a sink.

24   Arteries and veins are two types of blood vessel.
   a   List **three** differences between arteries and veins.
   b   Explain the reason for the differences.
25   Explain why the walls of capillaries are so thin.
26   Give an advantage of using artificial blood.
27   Stents are used to treat some circulatory problems. Describe how they work.
28   Describe the path taken by water molecules as they move through the plant in the transpiration stream.

Examination questions
Transport in animals and plants

1 Below is a drawing of human blood cells.

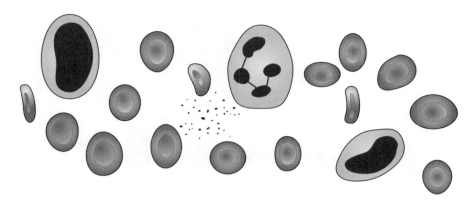

  a What is the function of the red blood cell?

....

....

(1 mark)

  b How do the platelets help prevent infection?

....

....

(1 mark)

  c How do the white blood cells help deal with infection?

....

....

....

(2 marks)
(Total marks: 4)

2 The heart pumps blood around the body. The heart contains valves. Sometimes the valves are damaged by disease, and fail to work. This may lead to heart failure or heart attacks.

  a What is the function of the valves in the heart?

....

....

....

(1 mark)

b One treatment for damaged valves, which has been developed since the 1950s, is to replace them with an artificial valve.

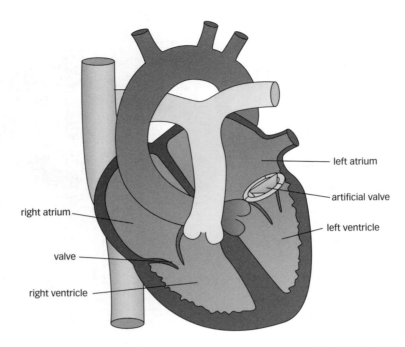

- Damaged heart valves can be replaced by open-heart surgery.
- Once fitted the valves last up to 30 years.
- Valves can cause damage to blood cells, which results in blood clotting, so patients must take anticlotting drugs for the rest of their lives.

Suggest advantages and disadvantages of treating patients with artificial valves.

....

....

....

....

....

....

....

....

(4 marks)
(Total marks: 5)

3 Look at the drawing of the human heart.

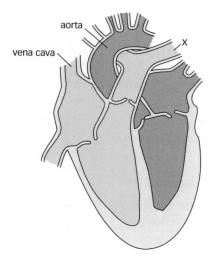

a What is the name of the blood vessel labelled X?

....

(1 mark)

b To which organ does blood vessel X take the blood?

....

(1 mark)

c Describe the route taken by the blood as it moves through the heart from the vena cava to the aorta.

....

....

....

....

....

....

....

....

....

....

....

....

....

(8 marks)
(Total marks: 10)

Research and development – ongoing review of ideas

In the exam you may be given information to read and then be asked to answer questions using that information and your own knowledge. The topics can vary but new technologies that can have a direct impact on our way of life are often chosen.

One such topic is the development of artificial aids to breathing.

Invention – the iron lungs of the 1920s

1　Read the information below. Using this and your own knowledge, answer the questions that follow.

We have learnt about the importance of ventilation to maintaining life. Before the 1920s any condition that paralysed the diaphragm and intercostal muscles would stop natural ventilation and the patient would die. This could be caused by exposure to some gases or the disease polio. In 1929 Dr Robert Henderson invented the iron lung to treat these patients.

Although the iron lung looks primitive, at the time of its invention it must have seemed like a miracle. Its first use was on an unconscious child, who responded rapidly to the treatment and woke up. By the 1950s when there was a polio epidemic, hospital wards filled with iron lungs were not uncommon, despite their high cost.

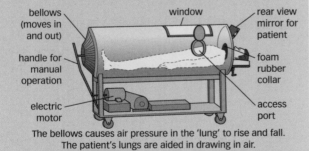

The bellows causes air pressure in the 'lung' to rise and fall. The patient's lungs are aided in drawing in air.

For the duration of their illness, the person is placed in a sealed chamber, with only their head exposed. Air is drawn out of the chamber, causing the pressure to drop in the chamber, and the chest to expand, drawing air into the lungs.

Skill – applying knowledge to an innovative solution

a　The iron lung replaces the actions of the body.
　i　Which parts of the body does it replicate?
　ii　Give one way they are similar and one way they are different.

Here you need to connect your existing knowledge to an unfamiliar situation. The connection is to explain how the body draws air into the lung. Then you explain that in the body the ribs and diaphragm carry out the function of the pressure change in the chamber. The similarities are often that the same outcome is achieved, but by a different mechanism.

Skill – evaluating the developments in science and technology

b　Suggest the advantages and disadvantages of treating patients with the iron lung.

When given a question that asks you to evaluate the advantages of a medical development, the advantage should often relate to prolonging life.

In contrast, when trying to list the disadvantages they often fall into one of two categories:

Social or financial
Here the text refers to the cost, and also that the body is enclosed in the machine, making it difficult for a doctor to treat the patient.

Ethical
This often means the quality of life; here it would be limited as patients would be in the machine for a long time.

Progression – modern ventilators

2　Now read the information below and answer the question that follows.

As iron lungs became more commonly used, doctors became frustrated by their limitations. By the 1950s they had developed a 'positive pressure ventilator'. This device did not place the patient in a large sealed chamber. Instead a pipe can be placed down into the lungs. Air is then gently pumped into the lungs. The pump uses positive pressure to expand the lungs. In long-term cases the pipe can be inserted into the windpipe, or trachea, at the neck.

Skill – Explaining technological developments

a　Explain how the positive pressure ventilators are a medical advancement.

Here you will be expected to apply your knowledge to this situation. You start from the problems identified for the iron lung, and look at the advancements implied in the text about the positive pressure ventilator. They include:
- allowing patients more freedom of movement
- allowing the doctor access to the patient's body for examination and surgery
- that they could be used to ventilate any patient during open-chest surgery, when the ribs don't work.

Moving forward – completing the story

Doctors are still working to improve the design of ventilators. Modern ventilators can detect whether the patient is making any attempt to breathe naturally. The device detects changes in pressure inside the lung caused by the patient, and responds by gradually phasing down its action, as the patient recovers. This would have been unthinkable in the iron lung of the last century.

AQA Upgrade

Answering questions that involve evaluation

QUESTION

In recent years scientists have developed artificial blood. This can be used in medical treatments requiring blood transfusions, but is particularly useful on battlefields. Below is a table of information comparing artificial blood to human blood.

Characteristic	Artificial blood	Human blood
Storage	room temperature	refrigerated
Shelf life	up to 36 months	up to 42 days
Active life	1–2 days	several weeks
Compatability	no antigens, so can be used in all patients	antigens present, needs tissue matching
Oxygen carriage	effective	decreases with storage time
Purity	pathogen free	needs to be screened for pathogens
Complications	increased risk of heart disease and heart attack	none

Suggest advantages and disadvantages of using artificial blood.

(5 marks)

G–E

Artificial blood is good because it lasts longer, but it gives you a heart attacke so its bad. Human blood has pathogens which kill us. But i also think it could be good for you because it lasts for several weeks, which is longer.

Examiner: Here the student has made several errors. The answer is not logical: it rambles and jumps from advantages to disadvantages. In the second half of the answer the student focuses on human blood, but the question asks about artificial blood; this might have caused the student's confusion. It is not clear whether points made are an advantage or disadvantage of artificial blood. Finally, none of the points have been explained. The student has simply re-written the points from the table.

D–C

The storage is up to 36 months, which is an advantage because it lasts longer. Also there are no disease causing pathogens in the artificial blood, like in the normal blood. So the patient who gets artificial blood won't catch any diseases like AIDS. The problems with artificial blood are that you can get ah heart attack. This would make you ill again after the artificial blood has been given to you. Also it doesn't last inside the patient, so they will need loads.

Examiner: This is a reasonable response, it is clear, and well organised, but there are two major problems. First, there are only four points in the answer. Look at the number of marks available. Always make one scientific point for each mark. The second problem is that two of the points have not been explained (storage and active-life points). They are just a repeat of the statements in the table. Repeating the source information is not enough. You must explain *why* you have chosen that point as an advantage or disadvantage. This is the evaluation of the evidence you have selected. 2 marks awarded.

B–A*

There are many advantages to using artificial blood. First the artificial blood has an increased shelf life and is easier to store. This means there is less waste of the blood. Secondly there are no antigens in the artificial blood which means that there will be no immune response or even death due to rejection of the blood. There are no pathogens in the artificial blood, which means there is less risk of infections like AIDS being transmitted to a patient. The disadvantages are that transfused artificial blood doesn't last long once given to the patient. So they need more transfusions. Last the increased risk of heart disease means that we may solve one problem but cause another. They may be ill in the future.

Examiner: This is a clear, logical, and well-argued response. Five points have been made. They are clearly explained, in order. The student has taken the facts and explained what the effect is, so it evaluates whether they are an advantage or disadvantage. Full marks.

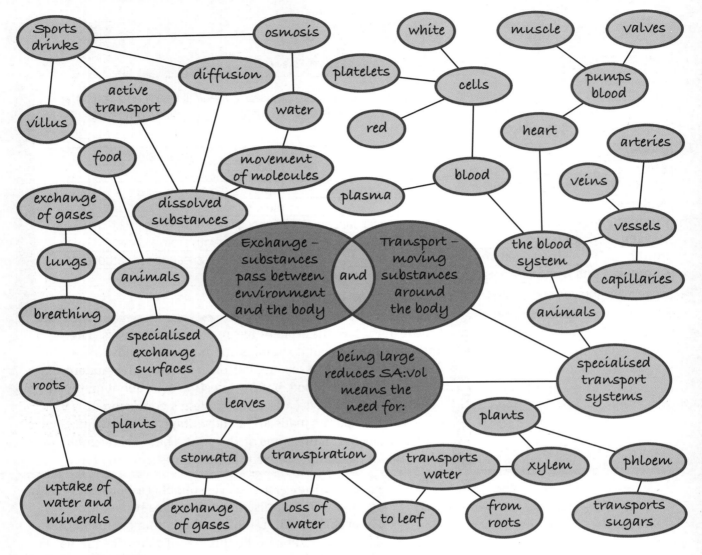

Revision checklist

- Dissolved substances can move into or out of cells by diffusion or active transport. Water moves into and out of cells by a special type of diffusion called osmosis.
- Keeping the body hydrated is important for human health. Water lost in sweat during exercise must be replaced.
- Sports drinks contain water for hydration, but also glucose for energy and ions to keep the body healthy.
- Active transport moves substances across cell membranes against a concentration gradient. This requires energy and a protein carrier.
- As organisms get bigger their surface-area-to-volume ratio gets smaller. This makes diffusion inefficient. To solve the problem they develop special exchange surfaces.
- All exchange surfaces are thin, have a large surface area, and have systems to maintain a concentration gradient.
- The villus is the site of absorption in the small intestine, and is efficiently designed.
- The lungs are the organs of gas exchange in humans. Breathing is the process which brings air in and out of the lungs.
- Transpiration is the loss of water from the leaves of the plant. Water leaves the plant through tiny pores called stomata on the under surface of the leaf.

- The rate of transpiration can be measured by a potometer, and can be affected by a number of environmental factors.
- A second problem for larger organisms is that they require a transport system to move substances around their body.
- Humans have a circulatory system that involves the heart, blood, and blood vessels.
- The heart is a muscular pump that pushes the blood around the body. Blood passes through the heart twice in one cycle of the body; it is a double circulatory system.
- The blood is the fluid that transports substances. It is composed of a liquid called plasma, and cells – red cells, white cells, and platelets – each with their own function.
- The blood is circulated in vessels: arteries take blood away from the heart, capillaries take blood through the tissues, where exchange occurs, and veins take blood back to the heart.
- Technology has developed artificial blood for use in transfusions, and stents to open blocked vessels. Artificial hearts or valves can be used to replace damaged ones.
- Plants also transport substances. The xylem transports water from roots to the leaves, whilst the phloem transports sugars from the leaf to other parts of the plant.

Keeping it balanced

The cells of our bodies are sensitive to any changes in the environment around them. It is essential to maintain a constant internal environment. If these conditions vary they are brought back to the body's normal value. This is called **homeostasis**.

Internal conditions that are kept within narrow limits are:
- pH
- **water** content
- ion (salt) content
- temperature
- blood **sugar** levels.

Internal condition	Effect on the body	Homeostatic function
pH	Chemicals like **carbon dioxide** or lactic acid lower the blood pH, damaging proteins.	The lungs control the levels by altering the breathing rate.
Water and ions	Water and ions are taken into the body in food and drink. If the content of these in the body is wrong, it affects osmosis. Water may move into or out of our cells, both resulting in damage.	The **kidneys** vary the levels in the urine. When we have drunk a lot of water, the kidneys produce lots of dilute urine, removing the water. When we are dehydrated the kidneys produce a little concentrated urine, thus saving water.
Temperature	Human body temperature is kept at 37 °C. It will rise and fall, which will affect cell function.	In the skin heat is lost or conserved.
Blood sugar levels	Sugar is taken in through our diet and used in respiration. Excess is stored. If the levels are too high or low then the body becomes ill.	The pancreas produces hormones that cause the sugar to be stored or released.

Many of the reactions in the cells of our body also produce toxic wastes, which must be removed. This process is called **excretion**.

Wastes that must be removed include:
- carbon dioxide
- **urea**.

Revision objectives

- understand that homeostasis is maintaining a constant internal environment
- understand that excretion is the removal of toxic substances from the body
- know how a healthy kidney produces urine

Student book references

3.15 Keeping internal conditions constant

3.16 The kidney

Specification key

✔ B3.3.1 a – c

Key words

homeostasis, water, sugar, carbon dioxide, kidney, excretion, urea, filtered, re-absorption

Waste	Production	Excretion function
Carbon dioxide	Carbon dioxide is a waste product of respiration. A build-up will lower pH, and damage proteins in the cell.	Breathed out through the lungs.
Urea	Produced in the liver by the breakdown of excess amino acids from proteins. This is toxic to cells because it is alkaline.	Removed by the kidneys, and stored in the bladder until urination.

The kidney

The kidneys are the organs that remove wastes like urea, excess water, and ions. This is done by producing urine in the kidney. This trickles down thin tubes and is stored in an organ called the bladder. When the bladder is full we urinate and release the urine to the outside of our body.

How does a kidney work?

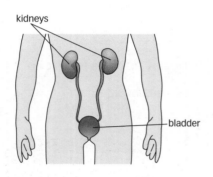
kidneys
bladder

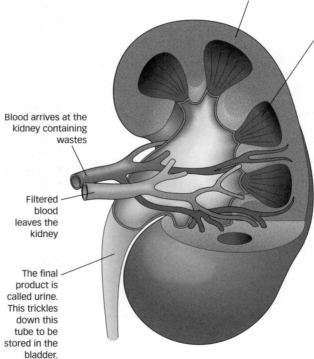

Zone of filtration – here blood is **filtered** and wastes like urea are removed from the blood. Unfortunately small useful substances are also removed. Large proteins and blood cells are too large to be filtered.

Zone of **re-absorption** – here the small useful molecules are taken back into the blood by active transport. The molecules include sugar, dissolved ions, and as much water as the body needs to maintain the correct level of hydration.

Blood arrives at the kidney containing wastes

Filtered blood leaves the kidney

The final product is called urine. This trickles down this tube to be stored in the bladder.

 The structure of the kidney.

Exam tip AQA

Students often muddle the terms homeostasis and excretion. Remember: homeostasis is about keeping conditions within narrow limits; excretion is about removal of wastes.

Questions

1   Why do we need to keep internal conditions constant?

2   What conditions does the body control?

3   **H** Explain how the kidney gets rid of waste.

Kidney failure

Kidneys are important organs because they remove toxic wastes from our bodies such as urea. If the kidneys stop filtering out these toxins, it can make us very ill. This is called kidney failure. This can happen in a number of ways.

Acute kidney failure – here the kidneys stop working suddenly as a result of disease or drugs, but will recover. Treatment is by **dialysis**.

Chronic kidney failure – here the kidneys gradually fail to work, and do not recover. This could result from overuse of a drug, diabetes, or genetic causes. Treatment is either by dialysis or transplant.

Kidney dialysis

The aim of dialysis is to remove the waste products from the blood, and restore the concentrations of all dissolved substances in the blood, like salts, to normal.

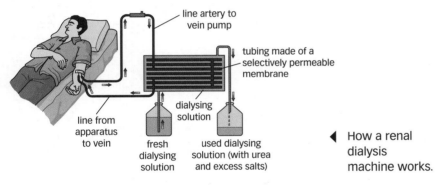

◀ How a renal dialysis machine works.

To achieve this, dialysis is usually carried out for five to six hours, three or four times a week.

The patient is attached to a dialysis machine, and their blood is taken from a vein and flows through the machine to be filtered.

In the dialysis machine the waste products are filtered out by diffusion.

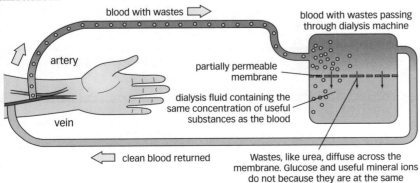

▲ How dialysis works.

Revision objectives

- ✔ understand how kidney dialysis works
- ✔ understand the process of kidney transplant and tissue typing
- ✔ be able to evaluate the two treatments

Student book references

3.17 Renal dialysis

3.18 Kidney transplants

Specification key

✔ B3.3.1 d – i

Kidney transplants

For patients with long-term kidney failure due to diseased kidneys, dialysis limits their quality of life. A transplant may be a better option. Here the diseased kidney is removed, and replaced with a healthy kidney. The healthy kidney is taken from a **donor** who might be a close relative or someone who has recently died. Care is taken to prevent **rejection** of the kidney by the **recipient's** immune system.

Organ rejection

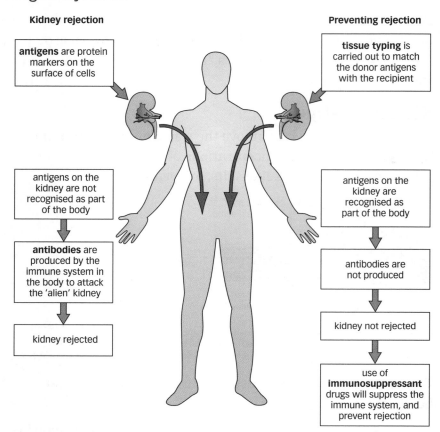

Kidney rejection

- **antigens** are protein markers on the surface of cells
- antigens on the kidney are not recognised as part of the body
- **antibodies** are produced by the immune system in the body to attack the 'alien' kidney
- kidney rejected

Preventing rejection

- **tissue typing** is carried out to match the donor antigens with the recipient
- antigens on the kidney are recognised as part of the body
- antibodies are not produced
- kidney not rejected
- use of **immunosuppressant** drugs will suppress the immune system, and prevent rejection

Evaluating treatments

Treatment	Advantages	Disadvantages
dialysis	• effective waste removal • allows time for the kidney to recover	• treatment time reduces quality of life • expensive
transplant	• long-term solution • better quality of life • cheaper in the long term	• tissue matching • lack of donors • rejection

Key words

dialysis, donor, recipient, antigen, antibody, tissue typing, rejection, immunosuppressant

Exam tip

AQA

You need to be able to *evaluate* the types of treatment. Remember this means discussing both the advantages and disadvantages of the treatments.

Questions

1 What is a kidney transplant?

2 How does kidney dialysis work?

3 **H** What causes organ rejection?

Body temperature

Our normal core body temperature is 37°C. Our body tries to keep its temperature at this level, or very close to it, all the time. This is called **thermoregulation**. 37°C is the temperature at which the body's organs work best. Sometimes our body will overheat or overcool depending on the environment or conditions in our body. Both situations are dangerous.

Overheating

- Causes – being in an environment with high external temperature, exercise, or dehydration, which prevents sweating.
- Dangerous level – any temperature above 40°C is dangerous.
- Body's response – increased **sweating** so more water needs to be taken in, looking flushed as more blood flows to the skin.
- Effect on the body – the higher temperature denatures enzymes, which harms cells.

Overcooling

- Causes – being in an environment with a low external temperature; babies have a large surface-area-to-volume ratio, and so lose heat, and the elderly can find it difficult to maintain heat.
- Dangerous level – any temperature below 35°C is dangerous, and is called hypothermia.
- Body's response – reduced sweating, **shivering** to generate heat, reduced **blood flow** to the skin, causing you to look paler.
- Effects on the body – the lower temperature slows enzyme reactions, making us unwell.

The thermoregulatory centre

The changes in the body temperature are detected in two ways:

1 An area in the brain called the **thermoregulatory centre** monitors the temperature of the blood flowing through the brain.
2 Nerves in the skin detect skin temperature and send messages to the thermoregulatory centre.

If the temperature is too high or too low, the thermoregulatory centre will then trigger the body's response processes.

Revision objectives

- ✔ be able to describe how thermoregulation is controlled
- ✔ explain the methods used in the body to bring the body temperature back to normal when it overheats
- ✔ explain the methods used in the body to bring the body temperature back to normal when it overcools

Student book references

3.19 Regulating body temperature – overheating
3.20 Regulating body temperature – overcooling

Specification key

✔ B3.3.2

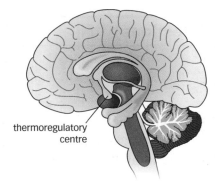

▲ The thermoregulatory centre (hypothalamus) in the brain.

thermoregulatory centre

Key words

thermoregulation,
thermoregulatory centre,
blood flow, sweating,
evaporate, shivering,
vasodilation, vasoconstriction

Mechanisms for thermoregulation

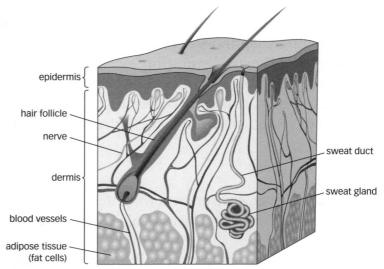

▲ Section through human skin showing sweat glands and sweat ducts.

Most of the mechanisms to help us control our body temperature involve the skin. It is important to be able to identify the major structures in the human skin.

Exam tip AQA

Students often muddle the terms they use in this topic. Remember when you talk about overheating, you also mean cooling the body down, and when you talk about overcooling, you are also talking about keeping heat in.

Questions

1 Name **one** thing that happens in the skin when we get too hot.

2 What is the thermoregulatory centre, and what does it do?

3 What is vasodilation?

When you overheat	When you overcool
AIM: to cool the body down	**AIM:** to keep heat in the body
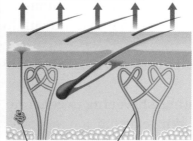	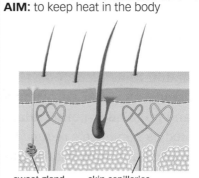
sweat gland skin capillaries	sweat gland skin capillaries
• Hair lie flat on skin; this reduces the layer of air, reducing insulation.	• Hairs stand up, trapping more insulating air.
H	
• Sweat glands release more sweat. This covers the skin and uses heat from the body to **evaporate** the water. • Muscles do not cause shivering.	• Sweat glands release less sweat, so less heat is lost by evaporation. • Muscles contract causing shivering. The contraction releases heat energy in respiration.
Vasodilation • Blood vessels in the surface of the skin expand or dilate. • This causes more blood to flow near the surface of the skin. • More heat is lost from the skin by radiation.	**Vasoconstriction** • Blood vessels in the surface of the skin get narrower, or constrict. • This causes less blood to flow near the surface of the skin. • Less heat is lost by radiation.

Regulating glucose in the blood

Glucose is an important molecule in our body. It is needed in cells to release energy during respiration. However, the level of glucose in the blood is usually kept within narrow limits. The normal level of glucose in the human blood is 72 mg per 100 cm³. If the level falls or rises, the body takes action to regulate the level.

How the body controls blood glucose

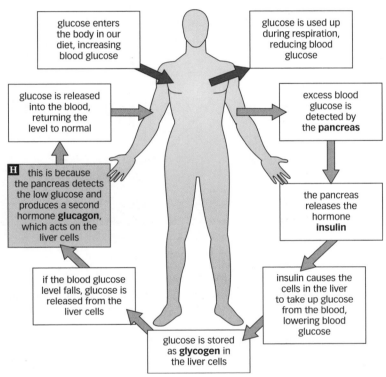

▲ Normal blood glucose level.

Type 1 diabetes

Type 1 **diabetes** is a condition where the pancreas does not produce enough insulin to allow us to control (by lowering) our blood glucose.

Key facts about type 1 diabetes

Frequency	approximately 1 in 800 people
Possible causes	genetic; viral; some drugs; trauma
Onset	usually between 10 and 18 years old
Symptoms	thirst, frequent urinating
How is it treated?	• careful control of diet, reducing intake of sugary foods • taking exercise • monitoring blood glucose levels • injecting insulin before a meal • regular health checks on circulation and eyesight

Revision objectives

- understand the roles of the pancreas, insulin, and glucagon in the control of blood glucose levels
- know the causes and treatments of type 1 diabetes

Student book references

3.21 Regulating blood glucose levels

3.22 Type 1 diabetes

Specification key

✔ B3.3.3

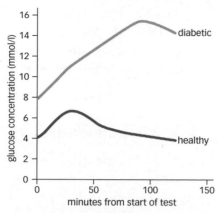

▲ Blood glucose concentrations in a normal and diabetic person immediately after a meal.

Questions

1 Name **two** hormones involved in the control of blood glucose.

2 Which hormone is not produced by a type 1 diabetic?

3 **H** Explain how a low blood glucose level is restored to normal.

Monitoring blood glucose levels

Blood glucose levels can be monitored by a simple digital recorder on a small sample of blood. This should be carried out at regular intervals. The graph on the left shows how the glucose levels would change after a meal, in both a normal and a diabetic patient.

Monitoring blood insulin levels

Insulin levels can also be monitored, but this is carried out in hospitals. Below is a graph that shows the insulin levels after a meal, in both a normal and a diabetic patient.

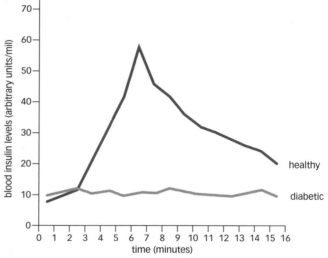

▲ Blood insulin concentrations in a normal and diabetic person immediately after a meal.

Improvements in modern treatments

- Modern sensors to monitor blood glucose are simple and more effective.
- Human insulin produced by genetic engineering is now used to inject, so there is less risk of allergies.
- Circulation is checked more thoroughly, which reduces complications due to diabetes, such as gangrene.
- Automated insulin pumps that release insulin into the blood via a catheter have been developed. This means no injections are needed, and the patient can live a more normal life.
- Research is being carried out to develop stem cells that might replace the damaged pancreatic cells.
- Gene therapy research aimed at stopping the damage being caused to the insulin-producing cells is underway.

Working to Grade E

1 Define homeostasis.
2 State one homeostatic function carried out by the following organs:
 a skin
 b kidney
 c pancreas
3 Define excretion.
4 What are the **two** major wastes produced in the body?
5 What is normal body temperature?
6 What is the function of the kidney?
7 Where are the kidneys located?
 • head
 • thorax
 • abdomen
8 What **two** types of treatment are used for kidney failure?
9 Name **two** causes of kidney failure.
10 What is the name given to the process of regulating body temperature?
11 Give **three** causes of the body overheating.
12 What is hypothermia?
13 Where is the thermoregulatory centre?
14 How does the body obtain glucose?
15 State **one** cause of type 1 diabetes.
16 What is glucose used for in the body?
17 Where is glucose stored in the body?
18 What is the effect of insulin on blood glucose levels?

Working to Grade C

19 Below is a diagram of the human kidney.

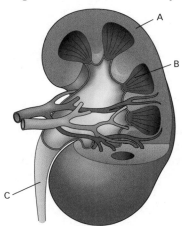

 a What happens in **A**?
 b Where does the waste product in **C** travel to?
20 Explain what causes the blood pH to fall.
21 Explain how urea is produced.
22 During dialysis describe what happens to the concentrations of urea in the blood.
23 What is an antigen?
24 During kidney transplant, who is the donor and who is the recipient?
25 The diagram shows the flow of blood through a dialysis machine.

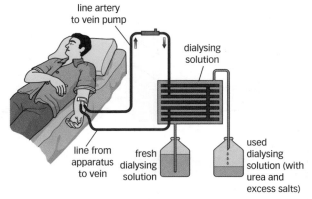

 a Draw an arrow labelled A to show the movement of urea during dialysis.
 b Label the partially permeable membrane.
 c Dialysis occurs frequently.
  i How long will a typical dialysis session last?
  ii Why does dialysis have to be repeated several times a week?
26 People often become flushed or red after doing exercise.
 a What causes this to happen?
 b What is the effect on the body?

27 Below is a drawing of the human skin.

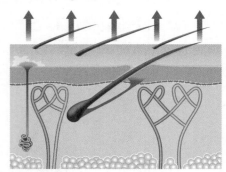

  a State whether the diagram shows the skin on a hot day or a cold day.
  b Give **three** reasons for your answer.
28 How does the body monitor the external temperature?
29 Why are body temperatures above 40 °C dangerous?
30 What regular healthcare checks are carried out on diabetics?
31 What problems occur if glucose levels become too high in the body?
32 Where is insulin made?
33 Does type 1 diabetes usually develop in young people or old people?
34 How has the development of an automated insulin pump improved quality of life for a diabetic?
35 What are the symptoms of type 1 diabetes?
36 Below are the results of regular blood glucose levels taken from two students, John and Peter.

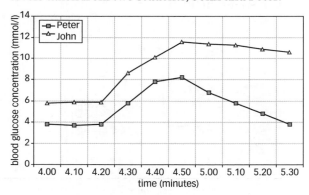

  a The two boys had a sugary meal.
    i At what time do you think they had the meal?
    ii Give a reason for your answer.
  b One of the boys is a diabetic.
    i Which boy is diabetic?
    ii Give a reason for your answer.
  c Describe what will happen to the insulin levels during this time, in the blood of:
    i John
    ii Peter.
  d What should the diabetic student do prior to a sugary meal?

37 Look back at the diagram in question 20. Identify **B** and explain how it ensures that useful molecules are not lost from the body.
38 Explain the effect of changes in the concentration of water or ions on the body.
39 During exercise the body produces more carbon dioxide. Explain the effect of this on the breathing rate.
40 Look back at the diagram in question 26.
  a Explain why the dialysis fluid is constantly being changed.
  b Explain how dialysis maintains and regulates the correct level of salts in a patient.
41 Explain the difference between acute and chronic kidney failure.
42 During kidney transplants organ rejection is a significant problem.
  a What is organ rejection?
  b How can we reduce the risk of organ rejection during transplants?
43 Patients will often prefer a kidney transplant over dialysis. What scientific argument could you use to back up this point of view?
44 Explain how sweating helps to regulate the body temperature.
45 Explain why we need to drink more fluid on a hot day.
46 Explain how hair could be used to increase **and** decrease the amount of insulation in an animal's body.
47 Explain what vasoconstriction is, and its effects on temperature regulation.
48 Explain the role of glucagon in regulating blood glucose levels.
49 Evaluate how modern developments have improved the treatment of diabetes.

Examination questions
Homeostasis

1 It is important for humans to maintain their water balance. Look at the diagram, which shows the input and output of water for a human in a typical day.

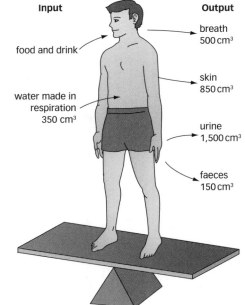

a To keep the water levels in the human in balance, what is the minimum amount of water this human would need to drink in a day?

....

....

.... cm³

(1 mark)

b Which of the processes shown in the diagram helps the body cool on a hot day?

....

....

(1 mark)

c Name another method the skin uses to cool the body.

....

(1 mark)

(Total marks: 3)

2 Below are the results of a blood sugar test for two students. It records the one and a half hours following their mid-morning break in school.

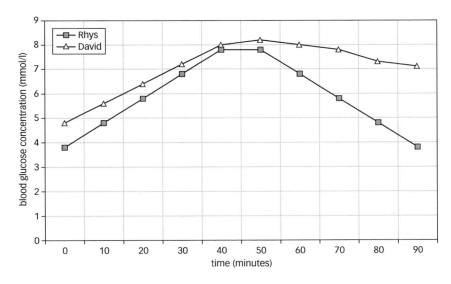

a What might have caused the blood sugar level to have risen in the blood of the two boys?

....

(1 mark)

b Which of the two boys do you think might be a diabetic? Explain why you think this.

Name of boy ....

Explanation ....

....

....

(3 marks)

c In a healthy person:

i What is the name of the hormone that causes the blood sugar level to return to normal?

....

ii What would happen to the amount of the hormone in a healthy person 30 minutes after a meal?

....

iii Which organ of the body produces the hormone?

....

iv Which organ of the body stores the sugar?

....

v What is the sugar stored as?

....

(5 marks)

d What causes the failure of control of blood sugar in a diabetic person?

....

....

....

(1 mark)
(Total marks: 10)

3 The kidney produces urine.

a What is the major waste product contained in urine?

....

(1 mark)

b Describe how the production of urine by the kidneys can help to regulate the water concentration of the blood.

....

....

....

....

....

(4 marks)

c When the kidneys fail, the patient needs treatment. There are two common treatments. Patients can have regular dialysis or they could have a kidney transplant.

The table below shows the concentrations of dissolved substances (measured in millimoles per litre, mmol/l) in the blood of a patient prior to dialysis and in the dialysis fluid being used during dialysis.

Substance	Concentration in mmol/l	
	Blood of kidney patient prior to dialysis	Kidney dialysis fluid
Glucose	5	5
Urea	25	0
Sodium ions	150	150
Chloride ions	150	150
Potassium ions	7	0

i The level of urea in a healthy person's blood is 5 mmol/l. What is the effect of this elevated value on a patient with kidney failure?

....

(1 mark)

ii Suggest what will happen to the concentrations of dissolved substances in the patient's blood during dialysis, and explain why.

....

....

....

....

....

(3 marks)

d Kidney transplants are often preferred by patients. However, there are problems of rejection. Explain what causes rejection of an organ transplant, and how the problem can be overcome.

....

....

....

....

....

....

....

....

....

....

....

....

....

....

(6 marks)
(Total marks: 15)

The human population

In the UK, as in most of the world, the human **population** has shown a massive increase in the past few hundred years. This rapid increase is due to:

- improved diet
- improved hygiene
- improved healthcare
- a reduction in the infant mortality rate.

Graphs of the human population of the UK show that the population has more than doubled in the past 100 years.

Sustainable living

As the population has increased in the past 100 years, the standard of living has also increased. There is a greater demand for manufactured products. This has placed a heavier demand on raw materials. The consequences of this are:

- the creation of more waste
- more land is used for building, farming, and extracting raw materials
- less fertile land is available for food production
- the world's resources are being used up faster than they can be replaced.

This lifestyle is now considered unsustainable.

Society has now recognised that they need to develop a more **sustainable** lifestyle. Sustainability is the use of resources without harming the environment by:

- replacing resources where possible, for example, replanting trees
- avoiding overuse of resources, for example, fishing quotas
- handling waste correctly to avoid **pollution**, such as recycling materials.

Human impact on the environment

Human activity has two major impacts. It reduces the land available, and it releases pollutants into the environment. Most of these impacts come from either agriculture or the development of towns and industries.

Loss of habitat

Humans need land for many reasons:

- building towns
- quarrying
- creating landfill waste dumps
- building industrial areas
- farming

Revision objectives

- know that the human population is increasing and that this increase is unsustainable
- understand how the actions of humans leads to pollution
- recognise the impact of various pollutants

Student book references

3.23 Human populations

3.24 Pollution

Specification key

- B3.4.1

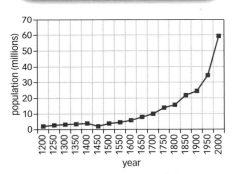

▲ The human population of the UK has been rising fast.

In doing these things, they destroy the natural environment. This usually means cutting down natural woodland. The woodland provides food and shelter to many species. Without these habitats there will be a reduction in biodiversity (number and types of organisms).

Pollution of land, air, and water

Pollutants are released into the environment. Different pollutants have different sources and different effects. Pollutants are released into the air, land, and water. The effects of many land pollutants occur when they are washed from the land into waterways.

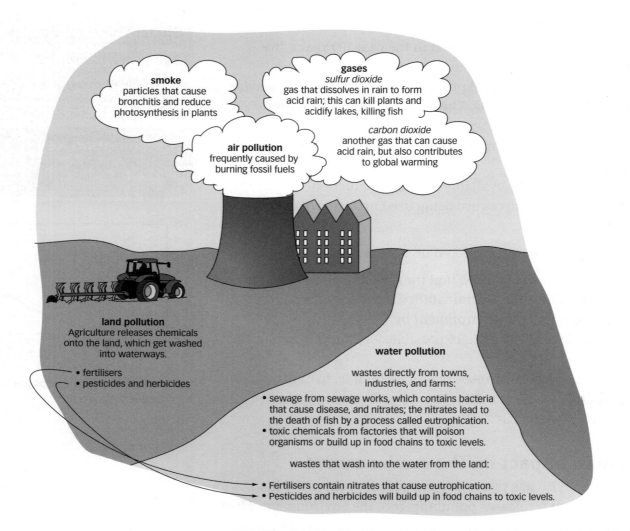

smoke
particles that cause bronchitis and reduce photosynthesis in plants

gases
sulfur dioxide
gas that dissolves in rain to form acid rain; this can kill plants and acidify lakes, killing fish

carbon dioxide
another gas that can cause acid rain, but also contributes to global warming

air pollution
frequently caused by burning fossil fuels

land pollution
Agriculture releases chemicals onto the land, which get washed into waterways.

• fertilisers
• pesticides and herbicides

water pollution

wastes directly from towns, industries, and farms:

• sewage from sewage works, which contains bacteria that cause disease, and nitrates; the nitrates lead to the death of fish by a process called eutrophication.
• toxic chemicals from factories that will poison organisms or build up in food chains to toxic levels.

wastes that wash into the water from the land:

• Fertilisers contain nitrates that cause eutrophication.
• Pesticides and herbicides will build up in food chains to toxic levels.

Questions

1   What is the link between the human population size and the levels of pollution?

2   What are the main pollutants in the air?

3   What is the effect of habitat destruction?

What is deforestation?

The large-scale felling of trees is called **deforestation**. This is happening worldwide, but is particularly common in countries like Brazil with the felling of tropical rainforests.

Why does deforestation occur?

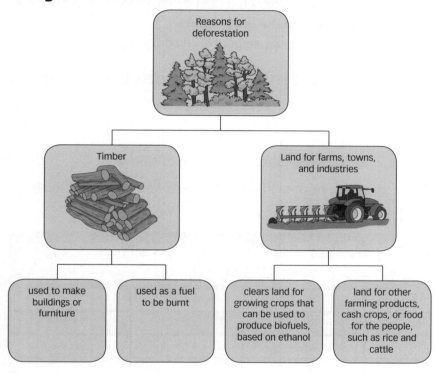

Revision objectives

- ✔ understand what deforestation is
- ✔ know the consequences of deforestation
- ✔ analyse and interpret environmental data

Student book references

3.25 Deforestation

Specification key

✔ B3.4.2

The cost of deforestation

Deforestation is responsible for a number of problems in the world. Some of the major problems are listed below.

Issue	Consequence
Slash and burn	Chopping down trees and burning the waste increases the amount of carbon dioxide in the atmosphere.
Effect on global gases	Deforestation leads to a rise in carbon dioxide levels in the atmosphere by: • the release of carbon dioxide during burning • the release of carbon dioxide during the decomposition of felled trees by microorganisms • a reduction in photosynthesis, so less carbon dioxide is taken up by trees and locked up in wood for many years.
Reduction in biodiversity	A diverse forest community is removed and replaced with a single crop. This provides few habitats and removes large amounts of the same mineral from the soil.
Build-up of methane	Cattle farms and rice fields both release methane into the atmosphere. This contributes to global warming.

Destruction of peat bogs

A second natural habitat that is being destroyed by humans is the **peat** bogs.

- Peat is produced over thousands of years by the preservation of moss in wet boggy areas that become acidic.
- One major use of peat was in the production of nutrient-rich composts for gardeners and plant growers.
- Unfortunately the extracted peat dries, decays, and releases carbon dioxide into the atmosphere.
- Many gardeners now use peat-free compost.

Analysing and interpreting environmental data

Frequently you will be asked to look at data on environmental topics and extract information from the data, or draw conclusions. Here is a simple example of this skill.

Below is a table of data about the extent and change of forest cover for a few countries.

Country	Forest area in 2005 (1000 ha)	Annual change in area 2000 (1000 ha)	Percentage (%) change in forest area in 2000	Annual change in area 2005 (1000 ha)	Percentage (%) change in forest area in 2005
Brazil	447 698	−2681	−0.62	−3103	−0.63
Cambodia	10 447	−140	−1.09	−219	−1.90
UK	2845	+18	+0.69	+10	+0.36
Norway	9387	+17	+0.19	+17	+0.18

▲ Data derived from forest resources assessment, 2005 (UN figures).

When you analyse the data, the figures indicate the following:
- Brazil and Cambodia are both experiencing deforestation – shown by the negative change in area.
- The UK and Norway are experiencing some reforestation – shown by the positive increase in area.
- The largest area of land being deforested is in Brazil – shown by the large value for the annual change in area.
- Cambodia has the quickest rate of loss – shown by the percentage-change figures.

Interpreting the meaning of this data suggests the following:
- Deforestation is a big issue in the tropical-rainforest areas of Brazil, as seen in the figures. This will result in the reduction of biodiversity and changes in global gas levels.
- This will have a greater effect in Brazil, because the area affected is greater.
- Reforestation is occurring in some countries, and this could be used as an example to other countries.

Questions

1   Give a reason why deforestation occurs.

2   How does deforestation affect biodiversity?

3   **H** Explain the relationship between deforestation and changes in global gas levels.

Global warming

Global warming is the overall increase in average global temperatures. Most scientists think that the increase in global temperatures is caused by heat being trapped in the Earth's atmosphere by a layer of gases called greenhouse gases.

Human activity is greatly increasing the amounts of these gases in the atmosphere. Scientists are worried that this is increasing the rate of global warming. The two major gases are:

*   carbon dioxide – released from burning fossil fuels, or from deforestation
*   methane – from cattle, rice fields, and decaying waste.

Revision objectives

*   understand the causes, process, and effects of global warming
*   know that biological materials can be used as a biofuel
*   understand the production, uses, and composition of biogas

Student book references

3.26 Global warming

3.27 Biofuels

Specification key

*   B3.4.3

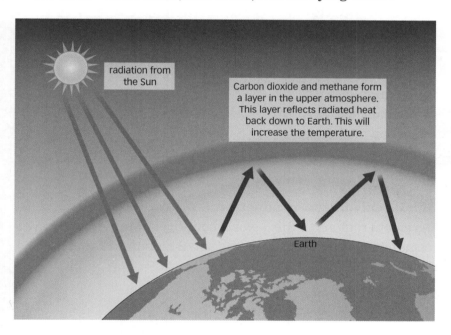

radiation from the Sun

Carbon dioxide and methane form a layer in the upper atmosphere. This layer reflects radiated heat back down to Earth. This will increase the temperature.

Earth

The effects of global warming

A rise in global temperature of only a few degrees can have a major impact on the Earth.

The effects of the oceans

Large bodies of water absorb large amounts of carbon dioxide. So oceans, lakes, and ponds remove carbon dioxide from the atmosphere, thus reducing its levels in the air. There are two ways this can happen:

*   Phytoplankton absorb carbon dioxide during photosynthesis.
*   Carbon dioxide dissolves in the water.

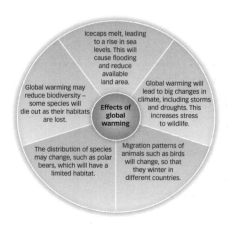

Key words

global warming, greenhouse gas, carbon dioxide, methane, biofuel, biogas

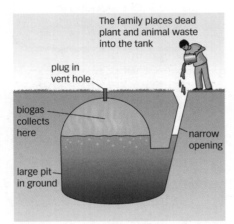

▲ A section through a biogas digester.

Questions

1   Name a biofuel.

2   How does the burning of fossil fuels cause global warming?

3   **H** Discuss the effects of global warming on wildlife.

Biofuels

Burning fossil fuels releases large amounts of carbon dioxide rapidly into the atmosphere, which has been locked up in the fuel for millions of years. Biofuels are a range of fuels from biological materials that are considered to be better for the environment. This is because the carbon dioxide produced when they burn is balanced by the carbon dioxide used in photosynthesis while the biological material is growing. So no overall increase in carbon dioxide in the atmosphere occurs. Biofuels include:

- wood
- alcohol
- biogas.

Biogas production

Biogas is made by the anaerobic fermentation of the carbohydrates in plant material and sewage by bacteria. Biogas is effectively a fuel produced from human waste. Biogas is a mixture of gases:

- methane (50–75%)
- carbon dioxide (25–50%)
- hydrogen, nitrogen, and hydrogen sulfide (less than 10%).

Biogas production can vary in scale.

Small-scale biogas production

In remote regions in third-world countries, families may have a small biogas digester to supply small amounts of fuel for cooking.

Large-scale biogas production

Here large commercial tanks are used. Again there is an anaerobic fermentation of waste, but the waste is constantly added. This method has widespread use, including at sewage works in the UK. The rate of gas production is affected by climatic conditions, mainly temperature, working best at 32–35 °C. Large volumes of the gas are produced, which are used to power vehicles, generate electricity, and heat homes.

Advantages of biofuels	Disadvantages of biofuels
Reduced fossil fuel consumption, by providing an alternative.	Causes habitat loss because large areas of land are needed to grow the plants.
No overall increase in levels of greenhouse gases, as the plants take in carbon dioxide to grow, and release it when burnt.	Habitat loss can lead to extinction of species.
Burning biogas and alcohol produces no particulates (smoke).	

Feeding the multitudes

As the human population has increased new methods have been sought to produce enough food to feed everyone. The food needs to be produced as locally to the population as possible. This will reduce transport costs and pollution.

Microbes as food

Fungi have always been used as a food source. By the 1960s scientists were becoming worried about the huge increase in the human population. If this increase continued, a consequence would be that there would not be enough protein to feed such a large population.

In 1967 a fungus was discovered called *Fusarium venenatum,* which could make a protein-based product. This protein is now called **mycoprotein**. This is produced in large **fermenters**.

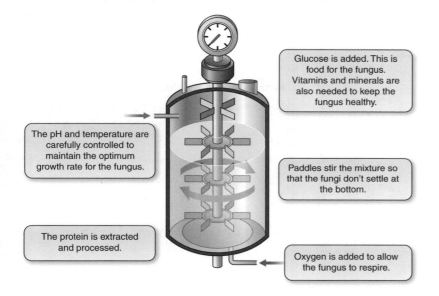

Glucose is added. This is food for the fungus. Vitamins and minerals are also needed to keep the fungus healthy.

The pH and temperature are carefully controlled to maintain the optimum growth rate for the fungus.

Paddles stir the mixture so that the fungi don't settle at the bottom.

The protein is extracted and processed.

Oxygen is added to allow the fungus to respire.

After production the protein is purified and made into many meat-substitute products.

Energy-efficient farming

Farming produces the food for the human food chain. The problem is that at every link in the food chain, energy and biomass are lost. This means that the longer the food chain, the more biomass and energy has been lost because of the greater number of links. So food from the end of a food chain is less energy efficient than food from earlier in the chain.

Revision objectives

- know that fungi are used to produce foods such as mycoprotein
- understand the methods of maximising efficiency of production of farmed foods
- know how fishing has become a sustainable method of food production

Student book references

3.28 Microbes and food production

3.29 Food chains and food production

3.30 Fishing

Specification key

- B3.4.4

Exam tip

For each of the processes listed on these pages make sure you can evaluate the positive and negative effects in terms of food production for an increasing population.

Advantages and disadvantages of factory-farming of animals

Advantages	Disadvantages
Less energy is lost in the food chain, so more is available for human consumption.	Greater risk of disease spreading through the animals as they are in close contact.
Less labour intensive, as animals are all contained in a limited area.	Some people feel that the technique is inhumane, or cruel to the animals.
Less risk of attack from predators like foxes.	Some people believe that the quality of the product is poorer.
Production costs are cheaper.	

Questions

1 What is mycoprotein, and what is it used for?

2 Discuss how efficient fishing techniques have caused problems.

Reducing energy loss

Biologists have been able to suggest a number of methods that can reduce energy loss in farming:

- eating vegetable products, for example, flour, rather than eating animals fed on vegetable products – this removes a link in the chain
- intensive **factory-farming** methods such as battery farming
  > reduces animal movement, reducing energy loss
  > keeps animals warm by rearing them indoors and in large numbers, so less energy is lost to the surroundings.

Fishing

Another common food is fish. As the population has increased, more fish have been caught from the world's oceans. Modern fishing fleets are very efficient at catching fish, using technology like sonar, and efficient, sophisticated nets. The impact of this is that the **fish stocks** in the oceans are in decline.

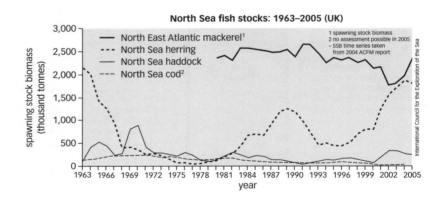

North Sea fish stocks: 1963–2005 (UK)

Protecting fish stocks

Governments became concerned about these low fish stocks. Numbers were so low that the species would soon be unavailable to be caught as food. The fish populations needed to be maintained at a sufficient size so that breeding continued successfully. The two main conservation methods employed were:

- net size – increasing the hole size allowed younger fish through; these could survive and breed
- fishing **quotas** – governments limited the numbers of fish that could be caught; this maintained a breeding population.

These measures continue and have allowed fishing to become an example of sustainable food production.

Working to Grade E

1 Define population.
2 Below is a graph of the human population of the UK.

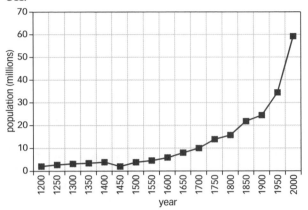

a What was the human population of the UK in 1700?
b How long did it take for the population of 1700 to double?
3 What is deforestation?
4 Give **two** uses of timber.
5 Peat bogs are habitats that are being destroyed by humans.
   a What do we use peat for?
   b Apart from loss of habitat, what other problem is created by the use of peat?
   c How can we overcome the problem?
6 What is global warming?
7 What is a biofuel?
8 Global warming is caused by greenhouse gases. What are the **two** main greenhouse gases?
9 What is the main constituent of biogas?
10 Below is a drawing of a small-scale biogas generator.

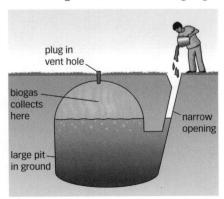

a What might people put into the generator to make the biogas?
b What do they mainly use the biogas for?
11 Global warming will have many effects, including affecting biodiversity. List **three** other effects of global warming.
12 What is a fermenter?
13 What is a fish quota?

Working to Grade C

14 What are the **two** major ways in which humans affect the environment?
15 Look at the graph in question 2.
   a Explain why there was little increase in the population of the UK between 1200 and 1400.
   b The population of the UK is increasing.
      i During which century is the rate of increase the greatest?
      ii Suggest **three** reasons that might explain the increase during that century.
16 Complete the table below, which shows some of the major pollutants humans release into the air.

Pollutant	Source	Effect on the environment
smoke	released from burning fossil fuels	
		contributes to global warming, and acid rain
sulfur dioxide		

17 We are being encouraged to live sustainably.
   a What is sustainability?
   b Suggest ways in which we can change our lifestyle to become sustainable.
18 List **three** examples of human activity that cause habitat loss.
19 Give **three** reasons why deforestation contributes to an increase in global carbon dioxide levels.
20 Below is a graph showing the rates of deforestation of primary forests (natural forest) in various countries.

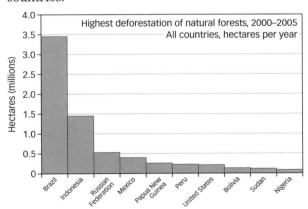

a The current annual rate of primary forest destruction in Brazil is 0.8% of the total forest. At this current rate, how long will it take for Brazil to completely lose its primary forests?
b Suggest reasons why Brazil might have undertaken such a high rate of deforestation.

21 Global warming is caused by greenhouse gases. What are the sources of the **two main** greenhouse gases?

22 Look back at the diagram in question 10.
   a What causes the biogas to be made inside the generator?
   b Why is this method of energy production particularly useful in remote parts of the world?

23 Global warming will have many effects. Explain how global warming will affect biodiversity.

24 What food source does the fungus *Fusarium* use in a fermenter?

25 Factory farming of animals is a common practice, for example, battery-farmed hens. Explain how battery farming of hens can reduce energy loss.

free range   battery

26 What has been the effect of increasing populations of humans on fish stocks?

27 Why is it more energy efficient to eat a producer than a carnivore?

28 Name **two** conditions that are controlled in a fermenter.

29 Give **two** methods by which trawler fleets became efficient.

30 Explain why scientists needed to find new sources of protein.

Working to Grade A*

31 Habitats are being lost as a result of human activity. Explain why loss of habitat is of such concern to biologists.

32 Farmers use a number of chemicals on the land that wash into the waterways.
   a Describe the effect of fertilisers in the water.
   b Explain how small amounts of pesticides washed into the water can become a problem.

33 Explain why it is important to treat sewage at sewage works before it is released into our streams and rivers.

34 Look back at the graph in question 20.
   a Explain the relationship between deforestation and biodiversity.
   b The 7th greatest loss of primary forest in the world is seen in the USA. This is a loss of 0.2% of the primary forests. However, they record a total gain in forests of 0.1% per year. Suggest a reason for these figures.

35 Look back at the diagram in question 10. Give **two** significant differences between a small-scale and a large-scale biogas generator.

36 Explain how the oceans help to reduce carbon dioxide levels in the atmosphere.

37 Explain why using biofuels is believed not to contribute to global warming.

38 Look back at the diagram in question 25. What is the difference between battery farming and free-range farming?

39 How has increasing the hole size in nets allowed fish stocks to recover?

40 The distance involved in transporting foods before they are consumed is called 'food miles'. Explain why scientists are so concerned about this issue.

41 Evaluate the pros and cons of factory farming.

1   Humans carry out a number of processes that have a harmful effect on the environment.

List A gives a list of such processes.

List B gives the effects of these processes on the environment.

Draw one line from each process in list A to the effect in list B.

List A
Process

List B
Effect on the environment

List A — Process	List B — Effect on the environment
	Contributes to global warming.
Burning fossil fuels releasing sulfur dioxide.	Particles blacken leaves and reduce photosynthesis.
Pesticides are used by farmers to kill pests.	Dissolves in rain to form acid rain.
Release of sewage.	Builds up in the food chain, killing other organisms.
Growing large areas of rice that releases methane.	Causes the death of fish by eutrophication.

(4 marks)
(Total marks: 4)

2   Some intensive farmers use a technique called battery or factory farming of animals like chickens to give improved yields.

Read the following points about factory farming.

> - Battery farming of chickens is where large numbers of hens are reared in cages.
> - This reduces the area of land needed, and fewer staff are employed to look after the birds.
> - The chickens are kept disease-free by treatment with antibiotics.
> - The birds are kept safe from predators because the cages are all kept indoors, in dark barns, using electric lights instead of daylight.
> - To protect the birds from scratching each other in the small cages they often have their claws removed.

a   Explain how the techniques used in factory farming of hens are a more energy-efficient way of farming.

....

....

....

....

....

(2 marks)

b   Some scientists have concerns about the use of battery or factory farming. Suggest reasons why.

....

....

....

....

....

....

....

(3 marks)

(Total marks: 5)

3   Biogas is an example of a biofuel. The drawing shows a biogas generator used in remote towns in countries like Nepal.

plug in
vent hole

biogas
collects
here

narrow
opening

large pit
in ground

a   Why is it important that air does not enter the generator?

....

....

....

....

(1 mark)

b   Why is it important that the generator is insulated in the ground?

....

....

....

....

(1 mark)

c   What are the advantages of using this type of generator in a rural community?

....

....

....

....

....

(2 marks)

(Total marks: 4)

Ethical issues raised by scientific developments

Research scientists are frequently developing theories and practices that have a direct impact on people's lives. For example, in medical research treatments are developed that can dramatically improve a patient's quality of life. On the face of it, medical developments like this seem to be for the good, and there would be no reason to think more about it. However, many medical developments can create a number of ethical, economic, or social issues. It may be the role of a clinical scientist to make decisions based on evidence to maximise the developments and save lives. Such an example can be seen in the treatments used for kidney failure.

Treating kidney failure

Patients with kidney failure have two major forms of treatment open to them:

- Kidney dialysis – This technique was first developed in 1943 by Dr Koff in the Netherlands during World War II. Since this time many improvements have been made to the technique. The process saves lives but requires several treatments a week, each for a few hours. However, this treatment does not provide a cure for chronic kidney failure.
- Kidney transplant – In 1954 Dr Joseph Murray performed a new treatment in Boston, US, where kidneys were transplanted from one individual to another. In general this treatment is superior, as it doesn't require regular, time-consuming dialysis sessions. People with transplants can lead normal lives post surgery.

It would seem that transplants are the perfect solution. However, there are a number of issues raised by the treatment.

The implications of kidney transplants

Cost

Transplant surgery is very expensive. In 2009 the typical cost was £17 000 per patient, with £5000 drug care per year. However, in the long term the surgical option might work out cheaper than dialysis, which costs £30 000 per year.

Number of donors

In 2009 there were 2497 kidney transplants in the UK. However, there were 6920 people still waiting for transplant. The need for kidneys outstrips availability.

Organ trade

In some countries people can sell their organs. However, the illegal sale of organs is an ethical issue in the UK.

Rejection

Tissue typing is needed in organ transplants. The closer the match, the more successful the anti-rejection (immunosuppressant) therapy.

Patient's state of health

Clinicians must evaluate a patient's state of health and age, to assess the likely success, and improvement to quality of life, of organ transplant. Some people raise concerns about transplanting into people with histories of long-term drug abuse.

Religious objections

Some religious beliefs prevent some patients having any form of transplant. Personal beliefs must be respected.

Reaching a decision

So, who gets the organ?

A clinical doctor must make a decision quickly. To do this they scrutinise the evidence available for each case.

Guidelines have been created by groups of doctors. These guidelines allow judgements to be made by individual transplant surgeons to decide which patient is best suited for a transplant. The decision uses social and scientific issues to select a patient who will gain the most, and thus there is less risk of wasting limited organs. The guidelines use scientific evidence and so prevent public or political disquiet.

AQA Upgrade

Dissecting exam questions

QUESTION

The skin plays a role in temperature control in the body. Look at the diagram of the skin.

1 Explain how sweating can regulate body temperature when the body overheats and overcools. *(3 marks)*

2 Explain **one** other way in which the structure of the skin is involved in temperature regulation. *(2 marks)*

epidermis
hair follicle
nerve
dermis
blood vessels
adipose tissue
(fat cells)
sweat duct
sweat gland

G–E

1 We sweat in the heat. This means we sweat more when its hot. This helps us in the heat.
2 When we get cold we shiver. This helps to warm us up because we are cold.

Examiner: This student has only told us one thing: that we sweat more in the heat. They have repeated the same point. The candidate has not dissected the question. They missed the command word, and have not explained the effect of sweating. They have also not seen the two sections – overheating and overcooling. 1 mark would be awarded.

This student has missed the instruction to identify a structure in the skin that helps temperature control. They did not need to have much additional knowledge as the diagram in the question lists structures. They have instead opted to talk about muscles. Therefore they gain no marks.

D–C

1 In the heat we produce lots more sweat. This is because it covers the skin and takes heat from the skin and evaporates. This cools our bodies. When we sweat on hot days this will make us thirsty.
2 The skin also has blood vessels that help to control the temperature.

Examiner: This student has picked up on the command word, and explained the effect of sweating. This has been done too much. The candidate has set out to say all they know about sweating. However, the candidate has missed the final section of the question, and not mentioned overcooling. The mark scheme will divide the marks up allotting some marks for both sections. Therefore, having missed the overcooling, they will gain only 2 marks.

This candidate has identified a structure in the skin that is involved in temperature regulation. But this time they have missed the command word to explain. Only 1 mark awarded.

B–A*

1 When we overheat, we will sweat more. This will cool our body down, by evaporating off the skin. But when we overcool, little or no sweat will be produced because we do not need to cool down.
2 There are blood vessels in the skin which are involved in temperature control. They can dilate when it is hot to lose heat, or constrict when it is cold to retain heat.

Examiner: This is a clear and logical answer. This question has three sections in it. First the command word 'explain' means tell the examiner how something happens. Secondly, the question asks about overheating and finally, overcooling. The candidate has dealt with both parts of the question, overheating and overcooling. They have noted the key word 'explain'. 3 marks awarded.

This is a well-structured answer. The question can again be dissected into parts. The command word 'explain' expects a statement of how the temperature is controlled. The next part of the question instructs the candidate to identify a structure in the skin involved in temperature control. Finally, the candidate should link that structure to its role in temperature regulation. This has been achieved here. 2 marks awarded.

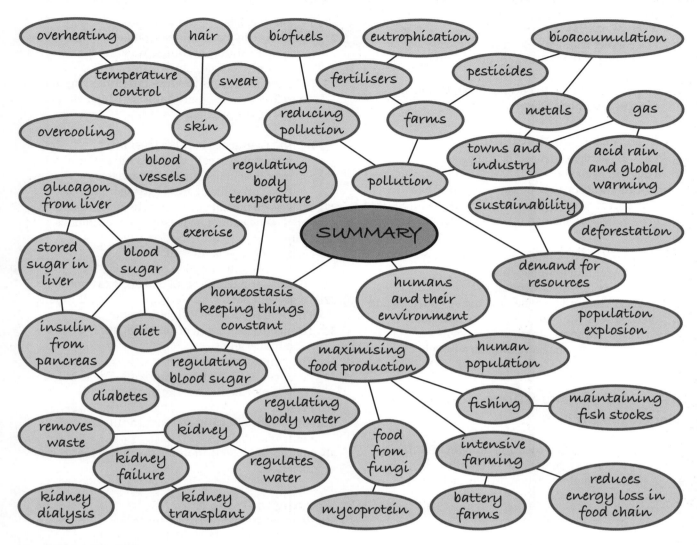

Revision checklist

- Animals need to regulate their internal environment by getting rid of wastes like urea and carbon dioxide, and maintaining constant internal conditions.
- Homeostasis is the process where animals maintain constant internal levels for body temperature, water, ions, and sugar.
- The kidney is an organ that filters wastes like urea from the blood and regulates the levels of water and ions in the blood.
- Kidney failure leads to a build-up of toxic wastes like urea, which can be fatal.
- Kidney failure can be treated by dialysis, where the blood is purified using a dialysis machine.
- Kidney transplants replace the damaged kidney but the transplanted kidney must be matched by tissue typing, to avoid rejection.
- The skin is the organ involved in temperature control.
- Overheating is prevented by sweating, hairs lying flat, and vasodilation.
- Overcooling is prevented by shivering, insulation, and vasoconstriction.
- Blood glucose levels are controlled by the hormones insulin and glucagon.

- Diabetes is a condition where insulin is not made, resulting in a failure to control the blood glucose level.
- The human population is increasing, which is putting extra demand on the Earth's resources.
- Humans release pollutants into the air and water from agriculture, towns, and industry.
- Deforestation is the process of removing large areas of natural woodland. This can result in an increase in carbon dioxide levels in the atmosphere, and loss of habitat.
- Carbon dioxide, and methane from rice fields and cattle, are greenhouse gases that lead to global warming.
- Biofuels are fuels from biological materials, which do not contribute to global warming.
- Humans are increasing food production to feed the increasing population by producing protein foods from fungi like mycoprotein.
- Farming methods can be made energy efficient by reducing energy losses in food chains.
- Overfishing had led to dwindling fish stocks. Governments introduced quotas and changes to nets to prevent overfishing.

Answers

B1 1: Diet, exercise, and health

1 The correct balance of foods and energy.
2 The rate at which cells carry out chemical reactions.
3 Cholesterol blocks blood vessels. This leads to heart attacks and strokes.

B1 2: Disease and medicine

1 Two of: bacteria, viruses, or fungi.
2 They kill bacteria.
3 Mutations occur in bacteria that make them resistant to existing antibiotics. So new antibiotics need to be developed to treat the infection.

B1 3: Immunity and immunisation

1 Phagocytes and lymphocytes.
2 The ability to resist infection.
3 Natural immunity is when the body is infected and then some of the lymphocytes are retained. Artificial immunity is when pathogens are introduced into the body causing a lymphocyte response.

B1 1–3 Levelled questions: Healthy living

Working to Grade E

1 Food such as sugar.
2 A diet that contains the right amount of different foods and the right amount of energy.
3 A person is malnourished if their diet is not balanced.
4 Obese
5 Energy
6 Energy
7 Build cells and repair tissue.
8 Any microbe that can cause an infectious disease.
9 A poison produced by a microorganism.
10 Bacteria
11 Either of: make us unwell or cause disease.
12 A drug that kills harmful bacteria in the body.
13 Paracetamol or codeine.
14 No
15 MRSA
16 Any three from: skin; mucus in the airways; stomach acid; blood clots as scabs.
17 a The lymphocyte is the cell with large rounded nucleus; the phagocyte has the lobed nucleus.
   b Ingests and digests pathogens.
18 Harmless (inactivated) or dead pathogen.
19 Measles, mumps, and rubella.

Working to Grade C

20 Amount of activity; proportion of muscle to fat; inherited factors.
21 In the cells.
22 More muscle and less fat results in higher metabolic rate.

23 Type II diabetes.
24 Hands spread many microorganisms because we touch many different things. Washing hands removes microorganisms.
25 Different pathogens produce different symptoms. Therefore the symptoms indicate which pathogen is causing the problem.
26 a C, B, A, D.
   b Virus attaches to cell; virus injects genetic material into cell; viral genetic material takes over host cell, causing it to make new viruses; host cell splits open, releasing new viruses.
27 Viruses do not carry out any chemical reactions, and so they cannot be poisoned by antibiotics.
28 Mutations
29 No living bacteria present.
30 To reduce contamination.
31 To transfer bacteria from one place to another.
32 To prevent bacteria entering or leaving the dishes. This will reduce the risk of infection.
33 To encourage growth of bacteria, but not the harmful varieties.
34 The immune system.
35 Produces antibodies or antitoxins. These kill the pathogen or neutralise the toxin.
36 Protein
37 Antibodies kill pathogens, and antitoxins neutralise toxins.
38 Making somebody immune to (able to resist) an infection.

Working to Grade A*

39 Reduces fat and increases muscle.
40 Cholesterol builds up in the walls of blood vessels. It can block the blood vessels, resulting in a heart attack.
41 Microorganisms multiply in the body and produce toxins that make us unwell.
42 Antibiotics kill many bacteria, but one or two may survive due to a mutation. The surviving bacteria will be resistant to the antibiotic and will multiply. They pass the resistance on to the next generation.
43 It produces bacteria that cannot be treated by antibiotics.
44 Use antibiotics less (for serious infections only) and ensure the full course of antibiotics is taken.
45 Each type of pathogen has a particular shaped antigen on its surface. The antibodies made by the lymphocyte have a specific shape that can only lock onto a particular shaped antigen.
46 A person receives a harmless or dead version of the pathogen. There is an immune response and the person's body makes antibodies. The body then has the ability to make these antibodies again rapidly if a real infection occurs. The antibodies destroy the pathogen before it can make the person ill.

B1 1–3 Examination questions: Healthy living

1 a White chocolate bar (1).
   b Carbohydrates and fats (1).
   c i   Raisin cereal bar (1).
      ii Because it contains the least fat (1) and the least sugar (1).
2 a Variable until about 1970, then a gradual decrease (1).
   b i   Between 1968 and 1970 (1).
      ii There was a sudden decrease in number of cases recorded after this time (1).
   c A dead or inactive virus is injected (1); this triggers lymphocytes to divide (1); some are retained (as memory cells) (1); these will respond/produce antibodies quicker if an active pathogen is introduced (1).

B1 4: Human control systems

1 Selecting the appropriate behaviour for a stimulus.
2 A receptor is a cell or organ that can detect stimuli.
3 Nervous systems act more quickly than hormonal systems. Hormonal responses last longer.

B1 5: Hormones and reproduction

1 The timing or the release of the egg, and the preparation of the womb for pregnancy.
2 To stimulate egg release, or to prevent egg release.
3 IVF is a medical procedure where fertilisation occurs in a test tube and the embryo is implanted into the female.

B1 4–5 Levelled questions: Controlling the human body

Working to Grade E

1 Any three of: light; sound; change in position; chemicals; pressure; pain; temperature.
2 Light
3 Skin
4 Nose and tongue
5 An electrical message that travels along a nerve.
6 a A: Nucleus; B: Insulating sheath; C: Nerve fibre; D: Synaptic end bulbs; E: Muscle.
   b Motor neurone
   c Impulse travels away from cell body toward the muscle.
7 A junction between two neurones.
8 Central nervous system and peripheral nervous system.
9 Knee jerk, hand withdrawal, or any sensible alternative.
10 Picking up an object, or any sensible alternative.
11 Glands
12 Kidney
13 37°C
14 The changes in the body that result in a person reaching sexual maturity.
15 Oestrogen and progesterone.
16 Luteinising hormone (LH).
17 The pituitary gland.
18 *In vitro* fertilisation.
19 The release of a mature egg from the ovary.

Working to Grade C

20 Takes impulses from receptors to the CNS.
21 Takes impulses from the CNS to the effector.
22 As a chemical message.
23 A rapid, protective, automatic response.
24 Stimulus → receptor → sensory neurone → relay neurone → motor neurone → effector → response.
25 Central nervous system.
26 Contracts
27 In the bloodstream.
28 In sweat, in urine, or when we breathe out.
29 Skin releases sweat, which evaporates, taking heat from the body.
30 Chemical reactions/muscle contraction.
31 Lost in sweat/in urine.
32 Source of energy.
33 Target organs
34 Maturation of the egg in the ovary.
35 Oestrogen and progesterone.
36 FSH and LH.
37 Blood clots (leading to heart attacks and strokes).
38 The part of the menstrual cycle when the wall of the womb is shed together with some blood.

Working to Grade A*

39 To maintain enzyme function.
40 Oestrogen and progesterone.
41 High levels of progesterone (low oestrogen).
42 To cause several eggs to mature and be released.
43 Families can control when they will start a family; it allows infertile couples to have children; embryos can be screened for disorders.

B1 4–5 Examination questions: Controlling the human body

1 Eye – Light (1); Ear – Sound (1); Tongue – Chemicals (1); Skin – Temperature (1).
2 Brain (1) and spinal cord (1).
3 a A chemical messenger made in one part of the body (1) that acts in another part of the body (the target organ) (1).
   b i   Folicle stimulating hormone/FSH (1).
      ii Luteinising hormone/LH (1).
   c They contain progesterone and oestrogen (1). These prevent the release of follicle stimulating hormone (FSH) (1), which prevents the maturation of eggs (1).
4 a i – B (1); ii – A (1); iii – C (1); iv – E (1).
   b Your answer should include six of these points in a logical order. 1 mark is awarded for each point up to a maximum of 6:
      • The sensory nerve ending detects the stimulus.

- An impulse is sent along the sensory neurone.
- Into the CNS/spinal cord.
- The impulse passes across a synapse.
- Into the relay neurone.
- The nerve impulse passes into a motor neurone.
- Passes out of the CNS/spinal cord.
- Impulse reaches a muscle.
- Arm is withdrawn.

   c Picking up a hot object (or any other logical alternative) (1).

5 a The reaction time gets quicker with more tests (1).

   b Eye (1).

   c i Used the computer to do the timing to milliseconds (1).

      ii The stimulus was repeated at least 10 times in each test (1).

   d i Her last result was slower than the previous value (1).

      ii Your answer could be either yes or no if it is well-reasoned: Yes, because the last does not follow the trend (it was significant); No, the last result was anomalous and not significant (1).

   e Select a group of at least five people who play computer games (1). Select a second group of five who do not play the computer games (1). Both groups carry out the test five times and in exactly the same way (1). Compare the results (1).

B1 6: Control in plants

1 A growth movement.

2 Hormones/auxin.

3 The auxin moves to the shaded side of the plant shoot.

B1 7: Use and abuse of drugs

1 They become addicted.

2 To give more reliable results.

3 Laboratory trials check that the drug is not toxic before it is given to a patient.

B1 6–7 Levelled questions: Control in plants and drugs

Working to Grade E

1 Light, gravity, and moisture.

2 Change in the environment.

3 Auxin

4 A chemical that affects our body chemistry.

5 Painkillers, antibiotics, and statins.

6 To ensure that the drugs work, and that they are not dangerous.

7 Tested on cells, tissues, and live animals.

8 Scientific testing of a new drug on human volunteers.

9 A dummy pill without the active drug.

10 a Sleeping pill, and to prevent morning sickness.

    b Treatment for leprosy.

11 Reduce blood cholesterol level.

12 a Taking drugs for no medical reason.

    b When a person needs a particular drug to maintain a functioning lifestyle.

    c Unpleasant sensations that occur when the body's chemical reactions do not function fully.

13 a To increase muscle development and to enhance performance.

    b Problems with reproductive cycle, and heart problems.

Working to Grade C

14 Growth movements called tropisms.

15 a Shoot should grow further up, roots should grow further down.

    b The shoot

    c Phototropism

16 a The shoot should turn up, the root will turn down.

    b Gravity

    c Gravitropism

17 Rooting hormones

18 a They avoid bias.

    b Neither the patients nor the doctors know who has the real drug.

19 An unwanted effect from a drug.

20 Cardiovascular diseases

21 a The greater the alcohol concentration, the greater the average braking time.

    b Alcohol slows people's reaction times.

    c Allows laws to be introduced to limit alcohol consumption for drivers.

    d Introduce a greater range of alcohol concentrations, and look for a point or concentration when the reaction time starts to decrease.

    e Have two drinks that look and taste the same; one will contain alcohol and one won't. Then repeat the tests.

22 They give the athletes an unfair advantage.

Working to Grade A*

23 Auxins cause plant cells in the shoot to elongate.

24 Auxins cause the shoots of broad-leafed plants to grow rapidly but not the roots, causing them to die.

25 The auxins move (by diffusion) to the shaded side of the shoot (away from the light). This causes the cells on the shaded side to elongate, which causes the shoot to bend toward the light.

26 a They will have grown straight up.

    b They all received equal light all round, therefore they do not bend.

    c The seedlings would begin to grow towards the light.

27 a No clear growth patterns.

    b No directional stimulus, such as gravity or light.

28 Shine light from one side only.

29 a Limb abnormalities in developing babies of pregnant women.
b They did not test the drug on enough types of animals or any pregnant women.
c Make sure the patient is not pregnant.
30 Tends to lead to people taking more powerful drugs, and can cause mental illness. However, it can be used to treat chronic pain.
31 No/disagree – because legal drugs are used by more people.
32 If they need to keep taking the drug. If they stop taking the drug to maintain a functioning lifestyle, they will suffer withdrawal symptoms.

B1 6–7 Examination questions: Control in plants and drugs

1 a A drug is a chemical that affects a person's body chemistry (1).
b Painkillers, antibiotics and statins (or any other sensible answer) (1).
c Alcohol, caffeine, nicotine, cannabis, steroids, cocaine and heroin (or any other sensible answer) (1).
d Most people use legal recreational drugs (1).
e Where people need a drug to maintain a functioning lifestyle. If they try to give up the drug, they suffer withdrawal symptoms (1).
2 a Gravitropism (1).
b They grow toward the light (1) and they can use the light for photosynthesis (1).
c Any four of the following points in a logical order. 1 mark will be awarded for each point up to a maximum of 4:
  • The plant produces a hormone called auxin.
  • In the shoot tips.
  • This diffuses back along the shoot.
  • The auxin moves away from the light to the shaded side of the shoot.
  • This causes the cells on the shaded side to elongate (get longer).
  • This bends the shoot toward the light.

How Science Works: Diet, exercise, hormones, genes, and drugs

1

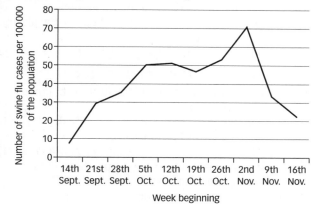

Number of swine flu cases per 100 000 of the population

Week beginning

1 mark will be awarded for equidistant spacing with ascending scales for both axes.
1 mark will be awarded for correct labels and units, and axes placed the correct way round.
2 marks will be awarded for accurate plotting of points (a tolerance of +/– 1 small square is allowed by the examiner). 1 mark will be deducted for one error; maximum deduction 2 marks.
1 mark will be awarded for joining the line from point to point.
2 The trend is upward until 2nd November – gains 1 mark. After 2nd November the trend falls – gains 1 mark.
3 Recognising either the slight rise on 5th October, or the slight dip on 19th October will gain 1 mark.
4 9th November as there the trend begins to fall.
5 10 cases (a tolerance of +/– 5 is allowed by the examiner).

B1 8: Adaptations of organisms

1 Any feature that aids survival and reproduction.
2 High temperatures and lack of water.
3 Cold. Adaptations include: thicker fur, smaller ears and nose, rounded body shape, insulating fat.

B1 9: Distribution in the environment

1 Where two or more groups of species compete with each other for the same resources.
2 Availability of resources such as food, light, carbon dioxide, predators, prey.
3 Most populations will reduce because pollutants or air affect the quality of the habitat, for example by changing pH levels or oxygen levels in water, or by the presence of toxic chemicals that poison the organisms. Some species (indicator species) are able to survive in such conditions and will increase.

B1 10: Energy and biomass

1 The mass of living material.
2 The total dry mass of all the individuals in the link of the food chain.
3 By keeping food chains short – fewer links mean smaller energy losses. Animals are kept warm and in small enclosures to reduce energy loss through movement and heat. By reducing pests, which compete with the organism for its energy.

B1 11: Natural cycles

1 The breakdown of the bodies, or wastes, of organisms.
2 The constant cycling of the element carbon between the living and the non-living world.
3 Detritivores eat bits of dead body such as dead leaves and digest them, releasing the elements in their wastes. This speeds up the process of decay.

B1 8–11 Levelled questions: Organisms and their environment

Working to Grade E

1 Fins/tail for swimming, gills for gas exchange.
2 Swollen stems to store water; leaves reduced to spines to reduce water loss; extensive root systems to absorb water.
3 High temperatures, but also extremes of salt and pressure.
4 Body shape designed to reduce water resistance to a minimum.
5 Light; space; water; minerals; carbon dioxide.
6 The release of harmful substances into the environment by humans.
7 Sulfur dioxide from burning fuels, or any other sensible answer.
8 Has a straw-like tail that can obtain oxygen from the air.
9 At the base.
10 Biomass
11 A link in the food chain.
12 An organism that feeds on dead organisms, such as an earthworm eating dead leaves.
13 Bacteria and fungi.
14 Respiration
15 Carbohydrates, fats, and proteins.
16 Via the food chain.

Working to Grade C

17 They can swim faster to catch prey or escape predators.
18 a Dive to great depths.
   b To find food.
19 Makes the body lighter for flight.
20 The colder the habitat, the thicker the fur, or vice versa.
21 a Plant A – the cactus lives in a hot dry environment; Plant B – the broad-leafed plant lives in a wetter environment.
   b Swollen stems; leaves reduced to spines; extensive root systems.
   c Swollen stems store water; leaves reduced to spines reduce water loss; extensive root systems absorb water.
22 Indicator species
23 a B
   b Oxygen is low because large numbers of bacteria/microorganisms respire, using up the oxygen.
24 Chemical energy in sugars.
25 Photosynthesis
26 The animals use less energy due to restricted movement and being kept warm due to heat loss.
27 Numbers of organisms at each tropic level and the typical mass of an organism at each level.
28 a Sketch should show: pyramid shape, three blocks, widest at base, narrowest at top.
   b Small numbers giving small biomass. Only a few foxes can be sustained by the number of rabbits in the habitat.
   c Heat; faeces; movement.

d The amount of energy falls at each link in the chain. There will not be sufficient energy to sustain too long a chain.
29 They must be equal.
30 Fossilisation: organisms die and are buried, or sink to the bottom of water; the organisms are covered by soils and exposed to high pressures; the bodies of the organisms turn into fossil fuels.
31 It releases large amounts of carbon dioxide into the atmosphere, at a faster rate that it is being removed from the atmosphere by plants.
32 a Wet and warm.
   b Microorganisms need both warmth and moisture to survive and function well. This condition was dry, so decay was slow.
   c Repeat the experiment several times.

Working to Grade A*

33 Plant B would die. This is because water loss would be more than water uptake.
34 a Thermophile bacteria
   b Because the human gut bacteria could not survive well at $50\,°C$.
35 Mayfly larvae are present because there is plenty of oxygen (and food), but there are few rat-tailed maggots as they are poor competitors.
36 The numbers of the two types of squirrels; the habitats in which they are both living; the types of food they both eat.
37 a The type of lichens in an area indicates the air pollution.
   b Answers should include reasons such as lichens are used because they live for a long time, and give an indication of the long term pollution levels.
38 a It is best to store bread in cool and dry conditions.
   b Conduct the experiment as above but use a range of different temperatures.
   c Any two from: the type of bread used; light; age of bread; amount of bread.
39 Trees will absorb carbon dioxide for photosynthesis, which will counter the amount of carbon dioxide being released by human activity.
40 When the animal dies, it decays; this is caused by microorganisms; the microorganisms respire, releasing carbon dioxide back into the atmosphere; the carbon dioxide is absorbed by the plant for photosynthesis; carbon compounds such as carbohydrates are made into the body of the plant.

B1 8–11 Examination questions: Organisms and their environment

1 a One from: grasping hands; tail for balance; grasping feet (1).
   b They feed at different times of the day (1) and they eat different foods (1).
   c They can make observations of similarities (1); organisms in the same group share many common features (1).
   d They share a common ancestor (1).

2	a	A – photosynthesis (1); B – combustion/burning (1); C – respiration (1).
	b	Bacteria (1) and fungi (1).
	c	1 mark per factor is awarded, up to a total of 2 marks. 1 mark per explanation is awarded up to a total of 2 marks. (Total question score: 4.) These could include:
		• 	Temperature – decay microorganisms are more active in warm (not hot) conditions.
		• 	Moisture – decay microorganisms are more active in moist conditions.
		• 	Oxygen – decay microorganisms are more active in aerobic conditions.
3	a	They are spreading out from the south east of the UK towards the north and west (1).
	b	They can fly long distances (1); they have a longer period of reproduction, so are active for longer in the year (1), and their numbers increase more so they can spread more per year (1).
	c	They compete with other ladybird species for food (aphids) (1); they eat other species of ladybird (1); they eat ladybird eggs (1).
	d	i	Large sample of results (1).
		ii	To make results accurate; it avoids mis-identification (1).

B1 12: Variation and reproduction
1	Variation
2	Genes, the environment, or a combination of both.
3	Sexual reproduction requires sex cells (gametes) produced by two parents to join; all offspring are genetically different from each parent. Asexual reproduction requires no fusion of gametes so only one parent is needed; all offspring are genetically identical to (clones of) the parent

B1 13: Engineering organisms
1	Two of: tissue culture, embryo transplants, adult cell cloning.
2	To make new products, e.g. hormones such as insulin, drugs, antibodies.
3	The process of transferring a gene from one organism to another. Ethical concerns include: superbugs, interfering with natural organisms.

B1 14: Genetically modified organisms
1	Crop plants that have had their genetic code altered by adding genes from other organisms.
2	They have an additional useful characteristic, such as pest resistance, herbicide resistance, increased vitamin content, or longer shelf life.
3	Any from: plants might escape and become more successful – superweeds; they might disrupt the natural food chain; uncertainty about the effects of eating GM crops; the seeds might be too expensive for some farmers.

B1 15: Evolution
1	The scientist who proposed the theory of evolution by natural selection.
2	The process by which those individuals with features best adapted to their environment (the 'fittest' animals in a generation) survive to reproduce and pass on their useful characteristics.
3	Darwin's theory suggests that variation between individuals occurs by chance (mutations) over several generations. Lamarck suggested that the changes occur in an individual of a species during its lifetime, rather than over several generations.

B1 12–15 Levelled questions: Genetics and evolution
Working to Grade E
1	Variation is the differences between individuals.
2	a	Genes
	b	Combination of genes and the environment.
	c	Genes
	d	Combination of genes and the environment.
3	Deoxyribonucleic acid – DNA.
4	The nucleus
5	Gametes/sex cells/eggs and sperm.
6	The process where two gametes meet and fuse.
7	Bacteria, plants, and many single-celled organisms.
8	A genetically identical organism.
9	Any one from: hormones such as insulin, drugs, antibodies.
10	Special enzymes
11	To include genes that kill insect pests.
12	No, for example the UK doesn't.
13	Any two arguments from: crops may escape, and out-compete wild flowers; disruption of food chains if insect-resistant plants are grown; uncertainty about the effects on human health; cost of testing, development and seeds; time taken for testing.
14	A chance change in an organism's genes.
15	Charles Darwin
16	Plants, animals, and microorganisms.
17	Plants: oak tree, daffodil; Animals: human, fish, crab; Microorganism: E. coli, salmonella.

Working to Grade C
18	a	D
	b	The manufacture of proteins, which affect how a cell works.
19	During reproduction. In sexual reproduction, genes are passed on in the eggs and sperm (or gametes).
20	Genes are passed to us from our parents: half from the father in the sperm cell, and half from the mother in the egg cell.
21	We are a mix of characteristics from our mother and our father, as the genes come from both parents.
22	All the new plants are identical, with the desired characteristics, and it is cheap.
23	B → D → A → C → E.
24	A host mother into which an embryo is implanted.

25 To produce large numbers of identical animals with the desired characteristics.

26 A sheep (Dolly).

27 They give an electric shock/use an electric current.

28 a i B; ii A; iii C.
    b Because it is a closer match, and there is no risk of disease transmission.

29 The crops can be sprayed with weedkiller, which will destroy the weeds but not the soya, and stop the competition from weed plants.

30 The process produces crops with the desired characteristic far quicker than selective breeding.

31 a It reduces the need for pesticides so there would be less pollution in the environment.
    b GM food has been eaten in some countries for 10 years with no ill-effects noted.
    c GM crops have a higher yield so can feed larger populations.

32 They make observations about the structure of the organisms and look for similarities and differences; organisms with more similarities are grouped together.

33 They share a common ancestor.

34 Humans and dogs share more similarities with each other than they do with fish, so humans and dogs have a closer common ancestor.

35 They live in similar environments, and share a feature which helps them survive in that environment. This is an ecological link.

36 Two from: the theory disagreed with religious ideas; there was not much evidence at the time; scientists didn't know about genes at the time, so couldn't explain any mechanism for evolution.

37 Because organisms cannot alter their genotype during their lifetime.

38 Over three billion years ago.

Working to Grade A*

39 The gene would code for the manufacture of a protein. This protein would have a certain pigmentation/colour. The cells in the (iris of the) eye would make large amounts of this protein.

40 Thousands of different proteins are needed to make a human body.

41 In asexual reproduction, only one parent is needed; there are no sex cells involved so no mixing of genetic information; all the offspring are genetically identical. In sexual reproduction, sex cells (gametes) combine from two parents (the mother and father); the offspring shares half its genetic material with the father and half with the mother; all offspring are genetically different from each parent.

42 Advantages include: quick, no need to find a mate. Disadvantages include lack of variation.

43 The new individuals are all genetically identical to, or clones of, the parent plant.

44 Any one from: all the offspring have the desired characteristic, large numbers can be produced.

45 The loop of DNA would be inserted into a fertilised sheep's egg.

46 a Variation in the peppered moth species: there is always a struggle to survive; the darker variation is more suited to the environment/it is 'fitter'; the darker variety survives and breeds; over time there will be more dark variety, as they are more successful.
    b It would be seen more easily by predators such as birds and eaten.
    c A mutation
    d The moth had an 'inner need' to change its body to a darker colour; it changed (so the genes must have changed); then the dark variety survived.
    e The environmental change happened quickly because their life cycle is short.

B1 12–15 Examination questions: Genetics and evolution

1 a Clones (1).
   b Grow on agar jelly (1), which contains plant hormones (1).
   c Any two from the following list – 1 mark for each: plants will have known characteristics, such as flower colour; all plants will be identical; the technique is faster.

2 Any four from the following list in a logical order, earning 1 mark each up to a maximum of 4:
   • Variation exists within the species.
   • Some variations are produced by mutations.
   • Some variations (long snouts) provide an advantage.
   • There is a struggle to survive.
   • The animals with the advantage are the 'fittest' and survive.
   • These pass the gene for the advantage (long snout) on to the next generation.
   • Gradually all animals tend to have long snouts.

3 5 marks available. 2 marks will be awarded for two positive points; 2 marks will be awarded for two negative points; the final mark will be awarded for a satisfactory evaluation of the pros and cons and a concluding judgement.
   Positives **include:** the technique produces more food, which can feed people in areas of famine; the increase in food production is because the food is engineered not to decay quickly, and is not affected by pests or pesticides; foods can be engineered to be enriched and more nutritious, e.g. with vitamins to improve health.
   Negative **include:** genes could escape, creating superweeds that are resistant to pesticides; the technology itself – any long-term health problems of the engineered plants are not known to the consumer; the process is expensive, and so some poorer countries (which have the greatest need to feed their population) might not be able to afford the seeds.

How Science Works: Surviving and changing in the environment

1   a   Accuracy is achieved by using a digital oxygen probe, which is very accurate; or counting the rat-tailed maggots in 1 square metre.
    b   Reliable results are achieved by doing repeat readings of oxygen levels to calculate an average. Alternatively, a large number of quadrat readings would ensure reliability.
2   More rat-tailed maggots are found at point A because of higher pollution; higher nitrate levels; which result in low oxygen levels. One mark for each of these three points.
3   a   Oxygen levels are low because the oxygen has been used up by aerobic organisms.
    b   Nitrates

B2 1: Cells and cell structure

1   An organelle/structure in the cell of plants and animals that contains the chromosomes. It controls the function of the cell.
2   Plant cells have: cell wall, cell membrane, cytoplasm, chloroplasts, vacuole, mitochondria, ribosomes, and nucleus. An animal cell does not have a cell wall, a vacuole, or chloroplasts.
3   Differentiation is the process whereby cells become specialised to do a particular job.

B2 2: Diffusion

1   Diffusion is the way molecules get into and out of cells, and this is necessary for cells to work.
2   Because particles don't move around in a solid.
3   The greater the surface area, the faster the rate of diffusion.

B2 1–2 Levelled questions: Cells and diffusion

Working to Grade E

1   a   i   A
        ii B
        iii C
    b   Chloroplasts, cell wall, and vacuole.
2   Inside the permanent vacuole.

Working to Grade C

3   a   Controls the activities of the cell.
    b   Supports the cell.
    c   Controls the movement of substances into and out of the cell.
    d   Traps light energy for photosynthesis.
4   a   Nerve cell – long extension; muscle cell – contractile proteins; palisade cell – chloroplasts.
    b   Nerve cell – takes impulses to the brain; muscle cell – contracts to allow the cell to shorten; palisade cell – allows photosynthesis.
5   Differentiation
6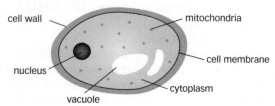

7   They don't have chloroplasts.
8   The net movement of particles from an area of high concentration to an area of low concentration, until the concentrations even out.
9   Oxygen
10 Distance, concentration gradient, and surface area.

Working to Grade A*

11 Make proteins
12 Cell wall is made of a different chemical; there is no distinct nucleus.
13 There is no diffusion occurring at A and C. At B, diffusion occurs into the cell. At D, diffusion occurs out of the cell.
14 This is the difference in concentration between the two areas.
15 a   This is how fast diffusion occurs.
    b   Distance – the shorter the distance the particles have to move, the faster the rate. Concentration gradient – the greater the difference in concentration, the faster the rate. Surface area – the greater the surface area, the faster the rate.
16 The lungs have a large surface area for the movement of oxygen. Therefore there is more surface area over which the oxygen molecules can move.

B2 1–2 Examination questions: Cells and diffusion

1   a   1 mark will be awarded for each correct structure/function completed in the table, up to a total of 8.

Structure	Function
Cell membrane	**Controls the movement of substances into and out of the cell.**
Nucleus	Controls the activities of the cell, contains DNA.
Cell wall	**Strong structure that supports the cell.**
Mitochondria	Releases energy from sugar during aerobic respiration.
Chloroplasts	**(Contains chlorophyll which traps light) to carry out photosynthesis.**
Vacuole	**Stores liquid (cell sap) used for support.**
Cytoplasm	Where many chemical reactions occur.
Ribosomes	Proteins are made here.

    b   1 mark will be awarded for each of the following three structures: chloroplast; cell wall; vacuole.
2   a   The net movement of particles (1) from an area of high concentration to an area of low concentration (1).
    b   Oxygen (1)

B2 3: Animal tissues and organs

1 When an organism is built of many cells.
2 Three from: mouth, oesophagus, stomach, small intestine, large intestine, pancreas, or liver.
3 A tissue is made of more or less similar cells working together, whereas an organ is a group of different tissues working together.

B2 4: Plant tissues and organs

1 Three from: epidermal tissue, mesophyll, xylem and phloem.
2 Phloem and xylem.
3 They are made of hollow cells with strong cell walls, which are stacked to form a long tube through the plant.

B2 3–4 Levelled questions: Tissues and organs

Working to Grade E

1 A group of similar cells working together.
2 a Muscle, glandular tissue, epithelial tissue, or any other reasonable answer.
  b Heart, stomach, or any other reasonable answer.
3 a Reproductive (A), circulatory (B), skeletal (C).
  b Heart
4 A group of different tissues working together at a specific function.
5 a Mouth
  b Large intestine
  c Small intestine
6 A: mouth, B: oesophagus, C: liver, D: stomach, E: pancreas, F: small intestine, G: large intestine, H: rectum.
7 a Supports the plant and transports molecules through the plant.
  b Production of food by photosynthesis.
  c Anchors the plant, uptake of water and minerals from the soil.

Working to Grade C

8 a Reproductive: reproduction; circulatory: transport; skeletal: support and movement.
  b As they are each made of a number of different organs working together at a specific function.
9 a It contracts causing the stomach to move and churn up the food.
  b Glandular tissue
  c On the outside and inside lining of the stomach.
10 a Produce a digestive juice (containing enzymes).
   b Produces bile, which aids digestion.
   c These produce a digestive juice, which is added into the mouth.
11 It is made of several organs working together at a specific function.
12 Epidermal tissue
13 Water

14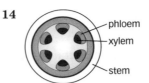

15 Stacked one above the other to form a long tube.
16 Inside the leaf.
17 Photosynthesis
18 Sugars

Working to Grade A*

19 Each tissue has a different role in the stomach. For the complete function of the stomach, all three roles are needed.
20 a The leaves
   b Other parts of the plant.
21 a Xylem
   b They have strong cell walls.

B2 3–4 Examination questions: Tissues and organs

1 a Answer should include any five of:
  - epithelium on the outside of stomach
  - covers the outside, protects the stomach
  - muscular tissue in wall
  - can contract and churn the contents up
  - glandular tissue on inside
  - produces acid and enzymes to help digest food
  - inner epithelium
  - to line the stomach.
  1 mark will be awarded for each up to a total of 5.

B2 5: Photosynthesis

1 To make food for the plant.
2 Light energy into chemical energy.
3 Water is absorbed from the soil by the roots, and it moves through the xylem to the leaf. Carbon dioxide diffuses through pores in the leaf and into the mesophyll cells.

B2 6: Rates of photosynthesis

1 Because there is more sunlight for photosynthesis.
2 They control the environment inside the greenhouse (levels of water, light, and carbon dioxide, and temperature).
3 Increase temperature, increase light, and increase carbon dioxide levels.

B2 7: Distribution of organisms

1 Place a transect line through the environment and use a quadrat at regular intervals to count the number of organisms.
2 Mean, median, and mode.
3 A combination of two things: the availability/ suitability of key factors such as temperature, light, carbon dioxide, nutrients, oxygen, and water; and the organism's adaptations to cope with the conditions.

B2 5–7 Levelled questions: Photosynthesis and distribution

Working to Grade E

1 Carbon dioxide and water.
2 Glucose/carbohydrates/sugars/starch and oxygen.
3 Sunlight
4 Chlorophyll
5 Leaf
6 Two from: plants, (some bacteria), and algae.
7 a A = upper epidermis; B = palisade mesophyll layer; C = spongy mesophyll layer; D = lower epidermis.
  b Palisade mesophyll layer.
8 The number of individuals of a species in a given area.
9 a In general, as the light increases the numbers of plants increase. But very high light levels will reduce plant numbers.
  b The more light, the more photosynthesis. But very high levels tend to increase temperature.
10 To count the numbers and types of organism in an area.
11 Polar bears live in colder environments.
12 All of the plant and animal populations in an area.

Working to Grade C

13 Carbon dioxide + water → glucose + oxygen
14 Through the stoma on the lower epidermis.
15 In a process that is controlled by a number of factors, the limiting factor is the factor which is at the lowest level and limits the rate of reaction.
16 The speed at which photosynthesis occurs.
17 a Carbon dioxide
  b No, it is now likely to be temperature.
18 a i Heat and carbon dioxide.
    ii They will increase the rate of photosynthesis (as long as there is plenty of light), and the glucose can be used for growth.
  b Stops the greenhouse getting too hot and killing the plants.
  c So that water can drain out of the pots to prevent rotting of roots.
19 It is very expensive, so reduces profit.
20 The temperature is hot in the day, and it is very dry.
21 a To make the results more reliable.
  b The readings were taken at set regular intervals that they could not change.
  c The number of daisies increases as you move further from the school.
  d The areas close to the school buildings have more shade.
22 a The bison eat the grassland plants, so high numbers of bison means a lower number of grass plants. So the bison need to move to an area with a fresh supply of grass plants.
  b The bison will either die or move on.

23 The mode is the most common value in a set of data. The median is the middle value when the data is in rank order.
24 a The larger the sample size, the more valid the results. This makes the data more accurate.
  b If results cannot be repeated then the conclusions may not be valid.

Working to Grade A*

25 a Arrow should pass though the stoma, through the air spaces up and into a palisade cell.
  b Sucrose is transported around the plant in the phloem. (Oxygen is released through the stoma.)
26 a i Nitrogen
    ii Nitrate ions from the soil.
  b Cellulose.
  c To store food and for growth.
  d Sucrose
27 Glucose
28 Because it is not soluble and so will not dissolve and leave the cell in water.
29 a A
  b As the light increases the rate of photosynthesis increases.
  c Another factor becomes limiting.
30 a On a warm day with plenty of sunshine and high traffic levels.
  b Because temperature and light levels are high, and the traffic gives off carbon dioxide, increasing its levels.
31 a Increased by electric lighting; decreased by netting or whitewash.
  b Maximum light will lead to a higher rate of photosynthesis. Too much light might raise the temperature too high and kill the plants.
32 Strawberries can be grown for a longer period during the year.
33 Light readings over a 24-hour period.

B2 5–7 Examination questions: Photosynthesis and distribution

1 1 mark will be awarded for each correct label, up to a total of 6.

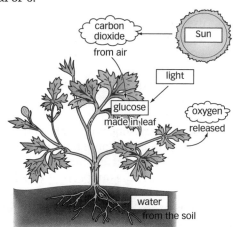

2 a You will gain 2 marks for plotting the points correctly and 1 mark for drawing the line correctly. 1 mark will be deducted for each plotting error up to a maximum of –2.

b 32.5 (+/–1) (1)

c 1 mark will be awarded for each of the following points, up to a maximum of 3.
- As the light intensity increases, the rate of photosynthesis increases.
- This is because more light energy can be trapped by the chlorophyll and used to build the sugars in photosynthesis.
- However, eventually there is no increase in the rate, despite the increase in light, probably due to other factors limiting the rate.

d 2 marks to be awarded as follows: When a process is affected by several factors, the one that is at the lowest level (1), will limit the rate of reaction (1).

e 1 mark for each of temperature and carbon dioxide.

3 a i Tube A (1)

ii 1 mark for each of the following parts of the answer, up to a total of 3:
- because the indicator has turned purple
- as the leaf has used up all of the carbon dioxide
- for photosynthesis.

b To keep the conditions the same in the two tests/to carry out a fair test. (1)

c Humans breathe out carbon dioxide (1), which will affect the indicator solution colour (1).

d i 1 mark for identifying that different amounts of carbon dioxide were absorbed (because the size of the leaf was different) and 1 mark for identifying that this was due to different amounts of photosynthesis (more carbon dioxide was absorbed by the bigger leaf in A, and more photosynthesis is able to occur in bigger leaves).

ii There would be no photosynthesis, as the leaf would be dead, so there would be no change in indicator colour. (1)

4 a Any 5 from the following list, up to a total of 5 marks:
- use a transect line
- lay it through the habitat, e.g. down a seashore
- place a quadrat
- at regular (predetermined) intervals
- identify the different seaweeds in each quadrat
- count the numbers of each type of seaweed in each quadrat (or estimate the % cover)
- plot the data as a graph.

b i By placing the quadrats at regular points along the transect. (1)

ii Repeat the experiment in the same location again. (1)

How Science Works: Cells and the growing plant

1 1 mark will be awarded for identifying that Dog's mercury will grow better at high light intensities. 1 mark will be awarded for a scientific explanation that where there is more light, there will be a higher rate of photosynthesis, making food for the plant to grow.

2 a 1 mark will be awarded for a biological variable from your hypothesis, such as the number of Dog's mercury plants or percentage cover.
1 mark will be awarded for the physical factor from your hypothesis, such as the light intensity.

b Refer to the guidance given in the Student Book.
1 mark will be awarded for outlining the correct use of apparatus; 1 mark will be awarded for a correct method of recording the data; 1 mark will be awarded for identifying factors to be controlled.

c 1 mark will be awarded for identifying that to avoid bias you should either place quadrats randomly or place them at regular, prefixed points.
1 mark will be awarded for identifying that reliable, reproducible results can be produced by having sufficient repeats, giving a large sample size, etc.

B2 8: Proteins and enzymes

1 They are built from long chains of amino acids, which are folded to give a specific shape.

2 Membrane proteins, hormones, antibodies, enzymes, and structural proteins.

3 They have an active site, which is complementary to the substrate. The substrate fits into the active site, and the reaction occurs.

B2 9: Enzymes and digestion

1 Digestion

2 Amylase breaks down starch into sugars.

3 The enzymes in the stomach work at an acid pH (pH2).

B2 8–9 Levelled questions: Enzymes and digestion

Working to Grade E

1 Amino acids

2 Diagram should be a long chain of beads, and each bead should represent an amino acid.

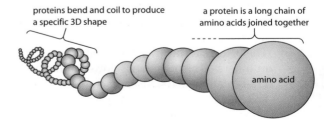

proteins bend and coil to produce a specific 3D shape

a protein is a long chain of amino acids joined together

amino acid

3 A biological catalyst that speeds up the rate of chemical reactions.
4 Proteins
5 The breakdown of large insoluble food molecules into small soluble food molecules.
6 a The liver
   b The gall bladder

Working to Grade C

7 The structure allows the specific shape to be formed. The shape is specific to their function

8

Type of protein	Function of protein
Antibodies	Bonds to a pathogen destroying them
Enzymes	Speed up the rate of chemical reactions
Membrane proteins	Allows substances into cells through membranes
Hormones	Controls the body's functions

9 Fibres in muscle cells.
10 c, a, d, b, e
11 a To break down large molecules into small ones.
    b To build large molecules from small ones.
12 pH
13 The reactions in the organism would be too slow for it to survive.

14

Region of enzyme action in the gut	Enzymes released	Reactions occurring
mouth	amylase	starch → sugars
stomach	protease	proteins → amino acids
small intestine	amylase	starch → sugars
	protease	proteins → amino acids
	lipase	lipids → fatty acids and glycerol (fats and oils)

15 It creates the correct pH for the stomach protease, and kills bacteria entering the gut.
16 It makes enzymes which digest foods in the small intestine.

Working to Grade A*

17 The shape of the active site of the enzyme is complementary to the substrate shape. No other substrate will fit into this active site.
18 a As the temperature increases, the rate of the reaction increases. It reaches an optimum temperature at which the rate is at its highest. Above the optimum temperature, the rate of reaction decreases rapidly, as the enzyme becomes denatured.
    b The molecules are moving faster as they have heat energy. They will then bump into each other more, increasing the rate at which they can react together.

c Above the optimum temperature, the increase in temperature begins to damage the shape of the enzyme. The shape of the active site is lost. The enzyme can no longer fit the substrate molecules.
19 To neutralise the acid from the stomach (emulsifies fats).

B2 8–9 Examination questions: Enzymes and digestion

1 a Enzymes speed up the rate of chemical reactions. (1)
   b i Stomach (1)
      ii 2 marks to be awarded as follows: Enzyme A works best in an acid pH (1) and the stomach has an acid pH (1).
      iii The line should continue down in a gradual slope, meeting the axis at about pH 8. (1)

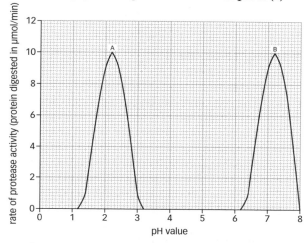

      iv 2. The examiner will allow a tolerance of +/− 0.2 in answers. (1)
   c 4 marks to be awarded as follows: a higher pH denatures the enzyme (1); this changes the shape of the enzyme (1); this damages the active site (1); so it is no longer complementary to the substrate (1).

B2 10: Applications of enzymes

1 To remove stains from clothes.
2 Proteases, carbohydrases and isomerases.
3 Above 40 °C the enzyme will become denatured and will no longer work. Below 40 °C the rate of reaction will become too slow.

B2 11: Respiration

1 Inside cells
2 Aerobic respiration requires oxygen; anaerobic respiration does not use oxygen.
3 Any two from: heart rate increases, which sends more blood to the tissues; breathing rate and depth increases, which increases oxygen uptake; glycogen is broken down, which releases more glucose to the cells.

B2 12: Cell division

1 Thread-like structures in the nucleus of every cell made of DNA, which contains the genes.

2 In areas of growth, repair, and asexual reproduction.

3 Mitosis produces identical daughter cells; meiosis produces cells with half the number of chromosomes.

B2 13: Stem cells

1 Unspecialised

2 Embryonic cells

3 One issue from:
- Embryonic: destroys an embryo; whether the benefits outweigh the costs; all life is valued.
- Umbilical cells: parents might have a second child for the wrong reasons.

B2 10–13 Levelled questions: Enzymes, respiration, and cell division

Working to Grade E

1 A cleaning agent.

2 Proteases and lipases.

3 a The reactions can be carried out at **lower** temperatures.

   b The reactions can be carried out at **lower** pressures.

   c The reactions will occur at a **higher** rate.

   d The cost of the process will be **lower**.

4 The release of energy from sugars in the cell.

5 Sugars such as glucose.

6 Building larger molecules; muscle contraction; maintaining body temperature.

7 The heart rate increases.

8 a Mitosis

   b Meiosis

   c Mitosis

9 Sperm are made in the testes of males, eggs are made in the ovaries of females.

Working to Grade C

10 Proteases digest proteins, lipases digest fats.

11 They remove stains that non-biological washing powders leave behind. They also remove stains at lower temperatures.

12 Most enzymes work best at 40 °C; above this they are denatured.

13 a Baby food:

   i Substrate is protein.

   ii Enzyme is a protease.

   iii Product: digested proteins/peptides/amino acids.

   Slimming bar:

   i Substrate is glucose.

   ii Enzyme is an isomerase.

   iii Product: fructose.

   b One from: they can't be used at high temperatures; they are expensive to produce.

14

	The reactants	The products
aerobic	glucose and oxygen	carbon dioxide and water
anaerobic	glucose	lactic acid

15 Mitosis

16 a 4

   b 2

   c 4

17 Any two from: embryonic stem cells; bone marrow; umbilical cord blood.

18 An undifferentiated cell (a cell that has not specialised as any one type).

19 Any one from: Parkinson's disease; spinal injuries; organ donation; diabetes.

20 Where cells become specialised to do a particular job.

Working to Grade A*

21 a Baby food: the protein is pre-digested, making it easy for the baby to digest and then absorb the products.

   b Slimming bar: fructose is sweeter than glucose, so less is needed in the slimming bar.

22 They digest cheap starch to produce sugar syrup, which is used in foods.

23 To pump the blood to the muscles quicker. This increases the supply of oxygen and glucose to muscles and takes away carbon dioxide.

24 When the muscles respire anaerobically during a shortage of oxygen, building up lactic acid. Oxygen debt is the amount of oxygen required to get rid of the lactic acid.

25 There is a drop in blood pH due to the build-up of carbon dioxide in the blood. This causes an increase in the breathing rate to increase the uptake of oxygen.

B2 10–13 Examination questions: Enzymes, respiration, and cell division

1 1 mark is awarded for each label, as shown.

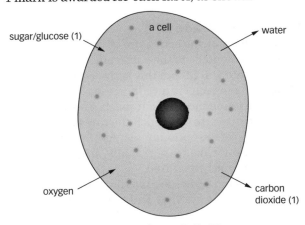

2 Gene, Chromosome, Nucleus, Cell. (3)

3 a Meiosis (1)
b 1 mark is awarded for a diagram completed as shown.

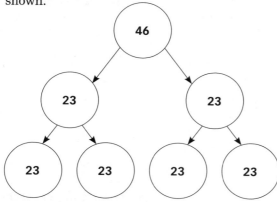

c Gametes (eggs or sperm). (1)
4 a Because they can develop into many other cell types in the body. These can be used in treatment. (1)
b 1 mark for any one from: embryos; umbilical cord blood; bone marrow.
c 1 mark for any one from: destruction of embryos, people having children to act as donors.

B2 14: Inheritance
1 Passing on characteristics from one generation to the next.
2 Gregor Mendel
3 Cross of two heterozygous individuals.

B2 15: Genes and genetic disorders
1 Which of the sex chromosomes we possess: XX = female, XY = male.
2 A recessive allele (c); a person must be homozygous cc to have the disorder.
3 It contains a genetic code. The code determines the order of the amino acids in a protein. This determines how the protein will work.

B2 16: Old and new species
1 Evidence of earlier life forms and how those forms might have changed.
2 Because the environment will change, and species will not be adapted to the new conditions and will die.
3 In the two populations of one species, different characteristics will be favoured in the different conditions. The individuals with these characteristics will survive. Their alleles are passed onto the next generation. Over time, each population becomes so different they cannot interbreed/form new species.

B2 14–16 Levelled questions: Inheritance and evolution
Working to Grade E
1 A section of DNA that controls a characteristic.
2 Mendel was the man who worked out the patterns of inheritance.
3 Pea plants
4 **a** XY
b XX
c Male
5 Deoxyribonucleic acid
6 A condition or illness caused by a defective gene.
7 The development of additional digits on the hands or feet.
8 The preserved remains of living things from years ago.
9 When a species dies out.

Working to Grade C
10 **a** Gametes: Ⓣ Ⓣ ⓣ ⓣ
b

Gametes	t	t
T	Tt	Tt
T	Tt	Tt

11 **a** When an allele always controls the development of the characteristic.
b When an allele will only control the development of the characteristic if the dominant allele is not present.
12 **a** 46
b 23
13 **a** Chromosomes present: XY × XX
Gametes: Ⓧ Ⓨ Ⓧ Ⓧ

Gametes	X	X
X	XX	XX
Y	XY	XY

b 50% or 1 in 2.
14 Answer should show the gene located in the same position on the second chromosome.
15 A spiral molecule like a twisted ladder, called a double helix.
16 The sequence of amino acids in a protein.
17 No
18 Identical twins
19 To identify a criminal from tissue evidence at the scene of a crime and to establish family connections such as paternity.
20 Soft bodies tend to rot quickly so there is not enough time for them to fossilise.
21 Some organisms are not well adapted to cope with the new conditions.
22 Speciation is where one species evolves into two new species.
23 It divides the population into two groups; each group then develops or evolves independently of the other.
24 There is very little evidence about the earliest life forms to help scientists explain the process.
25 Hunting

Working to Grade A*

26 **a** A list of the alleles present as a code.

    b A description of the characteristic in words.

    c The genotype has identical alleles.

27 Parents: Brown mouse × Brown mouse

    Genes present: Bb   ×   Bb

    Gametes: Ⓑ Ⓑ   Ⓑ Ⓑ

Gametes	B	b
B	BB	Bb
b	Bb	bb

    The chance of a white mouse is 1 in 4 or 25%.

28 **a** Inheritance was controlled by factors; factors are in pairs in adult cells; only one of each pair is in the gamete; offspring have two factors, one from each parent; the outcomes of crosses can be predicted.

    b There was a lack of scientific knowledge at the time – scientists had not discovered chromosomes.

    c He needed to control pollination, so that he knew which plants were the parents of which offspring.

29 Each new individual is a mix of chromosomes from the parents – half from the mother and half from the father.

30 **a** Three

    b Triplet

31 Each gene has a different sequence of bases. The order of the amino acids in a protein is determined by the sequence of bases in the DNA.

32 **a** Precious – Pp, Moses – pp

    b Parents: Moses   × Precious

    Phenotype: unaffected × affected

    Genotype: pp   ×   Pp

    Gametes: ⓟ ⓟ   Ⓟ ⓟ

Gametes	P	p
p	Pp	pp
p	Pp	pp

    The chances of having a fourth child with polydactyly is 50% or 1 in 2.

33 **a** When IVF is used to produce the embryos, they can be screened. This means that the cells of the embryo are checked for a specific allele.

    b It is discrimination against people with a genetic disorder; embryos with the disorder are discarded, which raises ethical concerns.

34 To avoid contaminating the sample with any other person's DNA.

35 **a** Bones

    b Lack of a complete fossil record.

    c Footprints in soft mud that fossilised.

    d Global temperature increase that the mammoth was unable to cope with.

36 • Isolation of two groups, one on each side of the river.

    • Variation in characteristics develops in each separated population.

• Different characteristics will be favoured in the different populations by natural selection.

• Over time, each group becomes different, until eventually they can no longer interbreed.

B2 14–16 Examination questions: Inheritance and evolution

1 Genes present: RR   ×   rr

   Genes in the gametes: R   ×   r

Gametes	R	R
r	Rr	Rr
r	Rr	Rr

    1 mark will be awarded for identifying each pair of genes, up to 2 marks.

    1 mark will be awarded for identifying each gene in the gametes, up to 2 marks

    1 mark will be awarded for correctly completing the table.

2 **a** Extinction is where a species has no individuals alive today. (1)

    b 1 mark will be awarded for each point below, up to a total of 3.

      • Restricted or limited diet – they only eat bamboo and they need a lot of it, also it dies back regularly.

      • Low birth rate, so not many new individuals produced to breed from, so low numbers in wild and small population.

      • Loss of habitat – humans destroy their habitat.

3 1 mark will be awarded for each point below, up to a total of 5.

    • The original population of squirrels becomes separated into two groups.

    • They become isolated from each other by the canyon.

    • Genetic variation occurs.

    • Natural selection occurs on both sides of the canyon.

    • Different features are favoured on each side of the canyon.

    • Over time, the two groups become so different they cannot interbreed.

B3 1: Dissolved substances

1 Water

2 Water, carbohydrate/sugar/glucose, ions, and caffeine.

3 Active transport requires carrier protein, osmosis does not.

    Active transport requires energy, osmosis does not.

    Active transport is against a concentration gradient, osmosis is not (or down a concentration gradient).

    Osmosis is the movement of water only.

B3 1 Levelled questions: Osmosis, sports drinks, and active transport

Working to Grade E

1 a Sweating – increases
   b Temperature – increases
   c Water intake – increases
2 Diffusion, osmosis, and active transport.
3 a Sweat more
   b Become dehydrated
   c Become more thirsty
4 High water concentration

Working to Grade C

5 Osmosis is a special kind of diffusion where water moves from an area of high water concentration to an area of low water concentration through a partially permeable membrane.
6 A membrane that allows some molecules through but not others.
7 a To the left.
   b Either: water concentration is higher on the right-hand side, or: sugar concentration is higher on the left.
8 Three from: to lubricate joints; to protect organs such as the brain; to carry substances around the body; to help regulate body temperature.
9 Ions
10

Ingredient	Function of ingredient
water	to hydrate the body
sugar/carbohydrate/glucose	a source of energy
ions	to keep muscles healthy
caffeine	to make us more alert

11 Sports drinks provide sugar/glucose/carbohydrate as a source of energy and ions to keep muscles healthy, whereas water does not contain these.

Working to Grade A*

12 It is too large to fit through the pores of the partially permeable membrane.
13 a It increases in size and mass. Cells become turgid (swollen).
    b There is no change in size or mass.
    c The chip decreases in size and mass. Cells become flaccid (soft).
14 a Sports lite
    b There are less dissolved substances in the drink.
    c It contains more glucose as an energy source.
    d Consuming too much sugar could cause a sudden surge in blood sugar levels (a sugar rush), triggering insulin release. This would then be followed by a sudden fall in blood sugar levels (a crash). This sudden change in blood sugar levels would cause problems for the body.
15 a Some dissolved molecules or ions move from a **low** concentration to a **high** concentration.

The movement is **against** a **concentration** gradient. This process is called active transport.
    b Uptake of ions by root hairs; movement of sodium ions out of nerve cells.
    c i Root hairs: nitrates, potassium, phosphates
           Nerves: sodium ions
       ii Root hairs: soil water
           Nerves: inside the nerve cell cytoplasm
       iii Root hairs: root hair cell cytoplasm and onto the xylem
           Nerves: outside the nerve cell
16 Carrier proteins in the membranes, and a source of energy (ATP).
17 Dead cells cannot make ATP or supply energy, which are necessary to active transport.

B3 1: Examination questions: Osmosis, sports drinks, and active transport

1 a The bag would have increased in size and mass. (1)
   b 1 mark awarded for each part of the following answer:
      • There is a concentration gradient between the solutions in the bag and the beaker.
      • Osmosis occurs, causing water to move into the bag of sugar solution (which has a lower water concentration).
      • After 30 minutes, there would therefore be a higher water concentration in the bag of sugar solution.
2 a 1 mark for each of the missing words, up to a total of 4.
      Sports drinks contain carbohydrates, water, and ions. Carbohydrates such as **glucose** are used in respiration to provide **energy**. Water is needed to **hydrate** the body. Mineral ions keep the **muscles** healthy.
   b 1 mark for identifying both of the best uses.

Drink contents	Hydration of the body	Energy supply for the body
Low sugar Dilute drink	✓	
High sugar Concentrated drink		✓

   c During sweating (1)

B3 2: Exchange surfaces

1 A specialised body surface over which molecules are efficiently exchanged.
2 Digested food such as glucose, amino acids, vitamins, and minerals.
3 Large surface area; thin surface; good blood supply; high turnover.

B3 3: Gaseous exchange in the lungs

1 To contract and move down to draw air into the lungs, and to relax and move up to expel air from the lungs.
2 Intercostal muscles
3 It diffuses across the alveolus wall, into the blood capillary, and finally into the red blood cell.

B3 4: Exchange in plants

1 They are the pores in the leaf through which gas exchange occurs. They also control water loss.
2 Any three from: light intensity; humidity; air movement; temperature

B3 2–4 Levelled questions: Exchange surfaces in the lungs and in plants

Working to Grade E

1 There is less surface area for diffusion to occur, so diffusion becomes inefficient.
2 a To make diffusion more effective.
   b Any three from: the digestive system; lungs or gills; plant leaves and roots.
3 In the chest/thorax.
4 The ribs protect the lungs and are used in the process of breathing.
5 Oxygen and carbon dioxide.
6 The ribs move down and in.
7 In the alveoli.
8 The diaphragm
9 The movement of air into and out of the lungs.
10 Mouth → trachea → bronchus → alveolus
11 A – ribs; B – diaphragm; C – lung; D – trachea; E – alveoli
12 Leaves and roots.
13 The movement of water from the roots to the leaves.
14 a Guard cells
    b i Arrow should show movement of oxygen out of the leaf through the stoma.
       ii Arrow should show movement of carbon dioxide into the plant through the stoma.
15 Potometer

Working to Grade C

16 a

Feature	How it improves diffusion
large surface area	provides more surface for greater diffusion
thin surface	provides a short diffusion distance
efficient blood supply	to maintain a concentration gradient

    b Any three from: dense capillary network; thin epithelium; presence of a lacteal; large surface area.
17 Surface area to volume ratio is much higher, which allows for efficient diffusion.

18 Any three from: large surface area; thin wall of the alveolus; good blood supply; lining of alveolus is moist.
19 The intercostal muscles relax, causing the ribs to fall down and in, and the diaphragm arches up. Both reduce the volume in the lungs, which increases the pressure.
20

Gas	Change
oxygen	decreased
carbon dioxide	increased
water vapour	increased
nitrogen	no change

21 Any three from: light intensity; temperature; air movement; humidity.
22 a Lower
    b Less heat from the sun to reduce water loss.
23 a The arrow should show water moving out.
    b During the day.
24 Larger surface area for exchange of gases. (An examiner would also accept that they can absorb more light.)
25 Increased surface area by the presence of root hairs.
26 It wilts and dies.

Working to Grade A*

27 Dense capillary network removes absorbed molecules, maintaining a concentration gradient. Thin epithelium provides short distance for molecules to have to move across.
Lacteal removes fats to maintain a concentration gradient.
Large surface area provides greater surface over which molecules can be absorbed.
28 a

Cell dimensions	Surface area	Volume	Surface-area-to-volume
3	54	27	54:27 2:1
6	216	216	216:216 1:1

    b The ratio gets smaller.
    c The surface area to volume ratio becomes too small, so diffusion becomes inefficient and it is necessary to have an exchange surface.
29 The intercostal muscles between the ribs contract, lifting the rib cage up and out, expanding the thorax.
The diaphragm contracts and flattens. This expands the thorax.
The volume inside the lungs increases, and the pressure decreases.
Air rushes into the lungs due to the low pressure.
30 To reduce water loss.
31 a Increasing the light will increase the rate of transpiration.

b Increasing the air movement will increase the rate of transpiration.

c Increasing the humidity will decrease the rate of transpiration.

B3 2–4 Examination questions: Exchange surfaces in the lungs and in plants

1 a 1 mark is awarded for the working, and 1 mark for the correct answer.
$320 \div 20 = 16$ breaths per minute

b It will increase. (1)

c i Oxygen (1)

ii 1 mark awarded for each of the following points, up to a total of 3:
- it has a large surface area
- it has a good blood supply
- it has a thin wall – only one cell thick.

2 a As the light intensity increases, the rate of water loss increases, but eventually there will be no further increase in water loss despite an increase in light intensity. (1)

b i They close. (1)

ii The rate of transpiration decreases with lower light intensity. (1)

c 2 marks awarded for explaining that the graph shows that high light levels (1) and windy conditions (1) increase the rate of transpiration (water loss), so the gardener will have to water their plants most on sunny and windy days.

B3 5: Circulation and the heart

1 To transport substances such as glucose and oxygen around the body.

2 They provide a short-term solution to heart failure that keeps patients alive while they await a heart transplant.

3 The muscle walls of the heart contracts. This forces blood through and out of the heart. The blood is pumped into arteries, which carry the blood around the body. Any backflow is prevented by valves.

B3 6: The blood and the vessels

1 Away from the heart.

2 It carries blood under higher pressure than in a vein.

3 White blood cells fight infection. Some engulf and digest microorganisms; others make antibodies that kill microorganisms. Platelets help to form blood clots, which prevent microorganisms entering the body.

B3 7: Transport in plants

1 In the xylem by the transpiration stream.

2 In the phloem by translocation.

3 The source (the leaf) to the sink (rest of plant, such as the root, seed, etc.).

B3 5–7 Levelled questions: Transport in animals and plants

Working to Grade E

1 Blood, heart, and blood vessels.

2 Oxygen

3 a A – right atrium; B – left atrium; C – left ventricle; D – right ventricle

b Blood vessel leaving left ventricle.

c Any of the valves between the atria and ventricles, or between the ventricles and the arteries.

4 Artery

5 Vein

6 Any two from: carbon dioxide, soluble products of digested foods, and urea.

7 a To transport oxygen around the body.

b Haemoglobin

8 Water and mineral ions.

9 Dissolved sugars

Working to Grade C

10 The blood passes through the heart twice per cycle around the body.

11 a Pulmonary artery

b Vena cava

12 Arrows should show blood flow from upper and lower vena cava into right atrium, down through valve into right ventricle, then up through valve and out via left and right pulmonary arteries.

13 To prevent the backflow of blood.

14 a The ventricles need thicker walls to produce a more powerful contraction, since they have to pump the blood out of the heart, while the atria only pump blood to the ventricles.

b It has to pump blood all round the body, while the left ventricle pumps blood to the lungs only.

15 They would fit an artificial valve.

16 a Any one from: keeps the patient alive; it is not rejected by the body.

b Any one from: it is a short-term solution; they often have wires that protrude through the skin.

17 a White blood cell.

b It helps form blood clots.

c It creates more room for haemoglobin.

18 Transport, protection, and regulation.

19 Any time when real blood is not available, for example during a war or major trauma.

20 A device to widen a blood vessel.

21 Either: xylem has dead cells, phloem has living cells, or: xylem has large angular cells, phloem has smaller cells.

22 The movement of dissolved sugars through the phloem from the leaves/site of photosynthesis (source) to the rest of the plant (sink), including growing regions and storage organs.

23 a A site of photosynthesis, where sugars are made.

b Leaf

c Growing regions and storage organs.

Working to Grade A*

24 a Any three from: arteries have a thicker wall than a vein; arteries have larger amounts of muscle; arteries have larger amounts of elastic fibres; arteries have narrower lumens; veins have valves while arteries do not.

b Arteries have a thicker wall than a vein because blood is under high pressure in the artery so a thick wall withstands the pressure. Arteries have larger amounts of muscle than veins as it allows the wall to withstand and maintain the high pressure.
Arteries have larger amounts of fibres than veins to allow stretch and recoil.
Arteries have narrower lumens than veins to help maintain pressure.
Veins have valves to prevent backflow of blood.

25 To allow substances to diffuse through the capillary wall quickly.

26 It replaces lost volume, and will not be rejected by the body.

27 The stent/a wire mesh is inserted into the narrowed region of the artery. The stent is opened/expanded using a small balloon, and this opens the artery to ensure proper blood flow.

28 Water molecules move into the root hair cell. They move from cell to cell across the root into the xylem. They travel up the xylem to the stem and leaves. From the leaf cells, water molecules move into the air spaces in the leaf, then out through the stoma as water vapour.

B3 5–7 Examination questions: Transport in animals and plants

1 a To carry oxygen around the body. (1)
b They form a clot at the site of a cut (a scab), which prevents microorganisms entering the body. (1)
c Phagocytes ingest/digest microorganisms. Lymphocytes make antibodies to kill microorganisms. 1 mark for identifying each type up to a total of 2.

2 a They prevent the backflow of blood. (1)
b 2 marks awarded for two different advantages; 2 marks awarded for two different disadvantages, up to a total of 4. For example:
Advantages: they extend lifespan; they improve quality of life.
Disadvantages: surgery carries with it a risk of infection; the valve can damage blood cells and this results in blood clots, which are a cause of heart attacks/strokes; the patient will have a lifelong dependency on anticlotting drugs, which are expensive.

3 a Pulmonary artery (1)
b The lungs (1)
c 8 marks available. Any of these points would be acceptable and earn 1 mark each (up to a total

of 8) as long as the full journey of the blood from the vena cava to the aorta is explained.
- Blood from the vena cava enters the right atrium.
- From the right atrium, the blood enters the right ventricle.
- Blood passes through the open valve.
- The right ventricle contracts, forcing the blood out.
- Blood enters the pulmonary artery.
- Blood travels to the lungs.
- Backflow of blood is prevented by the valve.
- Blood returns from the lungs in the pulmonary vein.
- Blood enters the left atrium.
- Blood passes into the left ventricle.
- The left ventricle contracts, forcing blood out into the aorta.

References to valves on both left and right sides will not be credited.

How Science Works: Exchange and transport

1 a i 1 mark each for diaphragm and intercostal muscles.
ii **Similarities:** 1 mark for either that they expand the chest or they cause air to be drawn into the lungs (ventilation).
Differences: 1 mark for explaining that muscles cause the chest to expand in the body, but changes in pressure carry out this function in the iron lung.
b 1 mark for each answer given, up to a total of 3.
Advantages: they keep patients alive while they recover (from diseases which paralyse the muscles).
Disadvantages: they are expensive; they are difficult for doctors to treat patients within; patients have a poor quality of life.

2 a 1 mark for each point up to a maximum of 3. They allow more freedom of movement/improve quality of life; they allow the doctors access to the body for examination and surgery; they can be used during open chest surgery when the ribs don't function.

B3 8: Homeostasis and the kidney

1 To maintain the normal functioning of the body cells.
2 Body temperature, blood glucose/sugar levels, water content, ion (salt) content, and pH.
3 Blood containing waste products arrives at the kidney; the blood is filtered in the outer zone of filtration; useful substances and water are re-absorbed in the inner zone of re-absorption; filtered blood leaves the kidney and waste passes down to the bladder in the urine.

B3 9: Kidney treatments

1 Replacing a damaged kidney with one from another person.
2 Blood is passed into a (dialysis) machine. Waste products are filtered out and the concentrations of all dissolved substances in the blood, such as salts, are restored to normal.
3 Antigens on the surface of the new organ are not recognised by the body. The body produces antibodies to attack the 'alien' organ and this leads to organ rejection.

B3 10: Controlling body temperature

1 Any one from: hairs flatten; sweat is released; vasodilation.
2 The thermoregulatory centre (hypothalamus) is an area of the brain. It monitors the temperature of blood flowing through the brain and receives messages from nerves about skin temperature. If temperature is too high or too low, it triggers the body's response to temperature change.
3 This is where the blood vessels in the surface of the skin expand or dilate. This causes more blood to flow to the surface of the skin. It results in greater heat being lost by radiation.

B3 11: Controlling blood sugar

1 Insulin and glucagon.
2 Insulin
3 Low blood glucose levels are detected by the pancreas. It stops making insulin, and starts to make glucagon. The glycogen in the liver is broken down into glucose. The glucose is released into the blood, restoring the blood glucose level to normal.

B3 8–11 Levelled questions: Homeostasis

Working to Grade E

1 Homeostasis is the maintaining of a constant internal environment.
2 a Temperature control
   b Water and ion control
   c Blood sugar control
3 The removal of waste products made in the body.
4 Carbon dioxide and urea.
5 37 °C
6 To remove wastes such as urea, excess water, and ions.
7 Abdomen
8 Dialysis and transplant.
9 Any two from: drugs; disease; diabetes; genetic causes.
10 Thermoregulation
11 Being in an environment with a high external temperature; exercise; dehydration.
12 When the body temperature drops below 35 °C.
13 It is an area of the brain/the hypothalamus.
14 Diet

15 Any one from: genetic; viral; some drugs; trauma.
16 As a source of energy.
17 Mainly in the liver.
18 It decreases blood glucose levels.

Working to Grade C

19 a Filtration of the blood. Waste products such as urea and small useful substances are removed.
    b The bladder
20 Carbon dioxide or lactic acid produced during respiration.
21 Produced when excess amino acids from proteins are broken down in the liver.
22 They go down.
23 Antigens are protein markers on the surface of cells.
24 The donor is the person who gives the kidney, the recipient is the person who receives the kidney.
25 a Arrow A should show movement from patient's arm, along line artery to vein pump, into tubes in dialysing solution, and out into container of used dialysing solution.
    b Label should be to the tubes inside the dialysis machine which are surrounded by dialysis fluid.
    c i 5–6 hours
       ii Wastes continually build up in the body and the patient would quickly get very ill if the toxins were not removed.
26 a Vasodilation: blood vessels in the surface of the skin expand. More blood flows to the skin.
    b More heat is lost from the skin by radiation.
27 a A hot day.
    b Any three from: hairs flattened; vasodilation/ blood vessels enlarged; sweat production; radiated heat from blood.
28 Nerves in the skin detect skin temperature.
29 The temperature denatures enzymes and this will harm cells.
30 Checks on circulation and eyesight.
31 Blood becomes more concentrated; water is withdrawn from cells by osmosis. Over the long-term, raised blood glucose is very damaging to a person's health.
32 The pancreas
33 Young people
34 No injections are needed and the patient can lead a more normal life.
35 Thirst and frequent urination.
36 a i After 20 minutes/around 4.20.
       ii This is when the level of blood glucose concentration starts to rise.
    b i John
       ii His glucose level is higher at the start. It rises higher, and is slower to return to normal.
    c i No change.
       ii Will rise after the meal, and then fall as the level of blood glucose returns to normal.
    d Inject insulin.

Working to Grade A*

37 The zone of re-absorption. Here, useful molecules (sugar, dissolved ions, water to ensure correct hydration) are reabsorbed by active transport.

38 It will affect osmosis, causing water to move into or out of cells, resulting in damage.

39 The breathing rate increases to remove carbon dioxide from the blood (via gaseous exchange in the lungs) more quickly.

40 a To maintain a concentration gradient in order to encourage diffusion of wastes out of the blood.

   b The dialysis fluid contains the correct levels of salts, and so there is no diffusion of salts out of the blood.

41 Acute failure is when the kidneys stop working suddenly.
   Chronic failure is when the kidneys gradually fail to work.

42 a When antibodies are produced by the immune system which attack and damage the new kidney.

   b Tissue typing prior to transplant; use of immunosuppressant drugs.

43 Dialysis limits the patient's quality of life, while a transplant will improve their quality of life. Transplants are cheaper in the long run, and a longer term solution.

44 When the body temperature rises, sweat is released from the sweat glands and covers the surface of the skin. Body heat is used to evaporate the sweat, and this heat loss cools the body and helps to regulate body temperature.

45 On a hot day, we sweat more to cool the body down. This means we need to drink more to replace lost water/ensure proper hydration.

46 When hairs stand up, they trap more insulating air and this keeps us warm. When hairs lie flat, they trap less air, reducing insulation. This cools the body.

47 During vasoconstriction, blood vessels in the surface of the skin get narrower. This means less blood flows near the surface of the skin, so less heat is lost by radiation. This helps to warm the body and maintain core body temperature.

48 When blood glucose levels becomes low, the pancreas detects the fall. It releases the hormone glucagon into the blood. Glucagon causes the liver to release glucose into the blood, which causes the blood glucose level to return to normal.

49 Modern sensors are simple and more effective, making it easier for diabetics to keep track of blood glucose levels.
   Genetically engineered human insulin poses less risk of allergies, making it safer.
   Thorough checking of circulation reduces complications that could cause serious illness.
   Automated insulin pumps allow patients to lead a more normal life.

B3 8–11 Examination questions: Homeostasis

1 a 2650 cm³ (1)
  b Sweating (1)
  c Vasodilation/hairs lie flat. (1)

2 a They might have eaten a sugary meal. (1)
  b David (1)
     His blood glucose level was higher before the meal. (1)
     When the level rises after eating a meal, it does not return to normal quickly. (1)
  c i Insulin (1)
    ii It would increase. (1)
    iii The pancreas (1)
    iv The liver (1)
    v Glycogen (1)
  d Their pancreas fails to produce enough insulin. (1)

3 a Urea (1)
  b 1 mark for each point, up to a total of 4:
    • When the body is dehydrated,
    • the kidney produces little and concentrated urine.
    • When the body is over hydrated,
    • the kidney produces large amounts of dilute urine.
  c i It is toxic to cells because it is alkaline. (1)
    ii Any three from the following list, earning 1 mark each up to a maximum of 3:
       • There will be no change in glucose, sodium and chloride ions
       • because the concentrations are the same in the blood and the fluid.
       • Urea and potassium ions will diffuse out
       • because there is a higher concentration in the blood than in the dialysis fluid.
  d 6 marks available. 3 marks will be awarded for three different causes, and 3 marks will be awarded for a sufficient explanation of how to overcome the problems identified.
     Rejection is caused by:
     • antigens on the surface of kidney cells
     • are not recognised as part of the body
     • antibodies are produced by the immune system
     • antibodies attack the alien kidney.
     To prevent rejection:
     • tissue typing is carried out
     • to match the donor antigens with the recipient's
     • antigens on the kidney are recognised as part of the body
     • no antibodies are produced
     • immunosuppressant drugs are used.

B3 12: Human populations and pollution

1 The larger the population, the greater the level of pollution.

2 Carbon dioxide, sulfur dioxide, smoke.

3 It removes food sources and shelter for many species. They can no longer survive. This reduces the biodiversity/number and types of organisms in the area.

B3 13: Deforestation

1 Any one from: timber for building, furniture, and fuel; to clear land to grow crops, e.g. biofuels, cash crops, and feed livestock, and to build farms, towns, and industries.

2 It reduces biodiversity.

3 Deforestation increases global levels of carbon dioxide, because there is less photosynthesis so less carbon dioxide is taken up by plants, and more decay and burning, both of which release carbon dioxide.

B3 14: Global warming and biofuels

1 Biogas, wood, or alcohol.

2 It releases greenhouse gases such as carbon dioxide. These trap heat in the atmosphere.

3 It causes loss of habitats, such as the melting of ice caps. This forces animals to change distribution and migration patterns. Global warming may reduce biodiversity – some species will die out as their habitats are lost. Climate changes such as droughts are stressful for animals.

B3 15: Food production

1 It is a protein-based food product made from a fungus. It is used as a food source in meat substitutes.

2 They have led to overfishing, resulting in falling fish stocks in the oceans.

B3 12–15 Levelled questions: Humans and their environment

Working to Grade E

1 The number of individuals of a species in an area.

2 **a** 10 million
   b Approximately 150 years.

3 Large scale felling of trees.

4 Used to make buildings/furniture or as a fuel.

5 **a** Nutrient-rich composts
   b Releases carbon dioxide as it decays.
   c Use peat-free compost.

6 The overall increase in average global temperatures.

7 A range of fuels made from biological materials.

8 Carbon dioxide and methane.

9 Methane

10 **a** Dead plant and animal waste.
   b It supplies fuel for cooking.

11 Any three from: habitat loss; species distribution changes; ice caps melt; climate change; migration patterns change.

12 A large tank for microorganisms to be grown in.

13 The number of fish it is permitted for a country to catch.

Working to Grade C

14 Humans reduce the amount of land available for other species and release pollutants.

15 **a** High mortality rate, especially among infants, due to poor hygiene and healthcare. Low numbers of individuals from which to breed. Food supply was limited and people ate a poor diet.
   b **i** 20th century
      ii Improved diet, hygiene, and healthcare.

16

Pollutant	Source	Effect on the environment
smoke	released from burning fossil fuels	**causes bronchitis, and reduces photosynthesis**
carbon dioxide	**released from burning fossil fuels**	contributes to global warming, and acid rain
sulfur dioxide	**released from burning fossil fuels**	**forms acid rain**

17 **a** Using resources without harming the environment.
   b Avoid overuse of resources; handle waste correctly; recycle materials; replace resources where possible.

18 Any three from: building towns and industrial areas; quarrying; farming; landfill waste sites.

19 Carbon dioxide is released during burning; carbon dioxide is released during the decay of felled trees; there is a reduction in photosynthesis because there are fewer trees to take up carbon dioxide.

20 **a** $100 \div 0.8 = 125$ years
   b Land is being used for farming, allowing cash crops to be grown for rapid income. Production of bioethanol.

21 Carbon dioxide from burning fossil fuels and deforestation; methane from cattle, rice fields, and decaying waste.

22 **a** Anaerobic fermentation of carbohydrates in plant material and sewage by bacteria.
   b It supplies energy in remote areas, where national grids cannot reach.

23 It reduces biodiversity by contributing to climate change that causes habitat loss.

24 Glucose

25 Less movement of animals, so less energy used. Less energy lost as heat, as being close together the surroundings become warmer.

26 Fish stocks have declined due to overfishing.

27 At every link in the food chain, more energy is lost. The longer the food chain, the more energy is lost. Therefore eating producers means less energy has been lost.

28 pH and temperature
29 Sonar and efficient, sophisticated nets.
30 There has been a huge increase in the human population, so we need to provide enough protein to feed everyone.

Working to Grade A*

31 It removes food and shelter for many species. It reduces biodiversity. It could cause many species to die out (become extinct).
32 **a** They contain nitrates, which kill fish by a process called eutrophication.
    b They build up in food chains to toxic levels.
33 Untreated sewage contains bacteria that can cause disease and nitrates, which can kill fish by eutropication.
34 **a** The greater the deforestation, the greater the loss of forest habitats, and the greater the reduction in biodiversity.
    b There is reforestation.
35 In a large-scale generator, waste is constantly added and much larger volumes of gas are produced.
36 They absorb large amounts of carbon dioxide as it dissolves in the water. Phytoplankton absorbs carbon dioxide during photosynthesis.
37 The plants take in carbon dioxide to grow and release it again when burnt. So there is no overall increase in carbon dioxide levels.
38 In battery farming, the chickens are reared in small cages, whereas in free range farming, the animals are allowed to roam freely.
39 It allows smaller/younger fish to escape so they can survive and breed.
40 The greater the amount of transport involved, the more fuel is burnt. This releases pollutant gases such as carbon dioxide into the atmosphere, which contributes to global warming.

41

Pros	Cons
Less energy is lost in the food chain, so more is available for human consumption.	Greater risk of disease spreading through the animals as they are in close contact.
Less labour intensive, as animals are all contained in a limited area.	Some people feel that the technique is inhumane, or cruel to the animals.
Less risk of attack from predators such as foxes.	Some people believe that the quality of the product is poorer.
Production costs are cheaper.	

In summary, the pros lead to cheaper, more plentiful food. The cons are inhumane treatment of animals, and the quality of the food is not as good.

B3 12–15 Examination questions: Humans and their environment

1 1 mark will be awarded for each correct match, up to a maximum of 4.

Burning fossil fuels releasing sulfur dioxide.	Dissolves in rain to form acid rain.
Pesticides are used by farmers to kill pests.	Builds up in the food chain, killing other organisms.
Release of sewage.	Causes the death of fish by eutrophication.
Growing large areas of rice which release methane.	Contributes to global warming.

2 **a** Lack of movement – uses less energy. (1)
       Birds lose less energy as heat – if they are close together they can share body heat. (1)
    b Any three from the following list earning 1 mark each up to a total of 3:
       • it causes behavioural problems such as scratching
       • diseases spread rapidly through the chickens, as they are so close together
       • there is an increased use of antibiotics
       • there are ethical issues about the humane treatment of animals.
3 **a** It can become explosive. (1)
    b It keeps it warm. (1)
    c Any two from the following list, earning 1 mark each up to a total of 2:
       • it provides fuel for cooking and heating
       • rural communities cannot always be on mains gas supply
       • it disposes of waste
       • it reduces use of fossil fuels.

Appendices

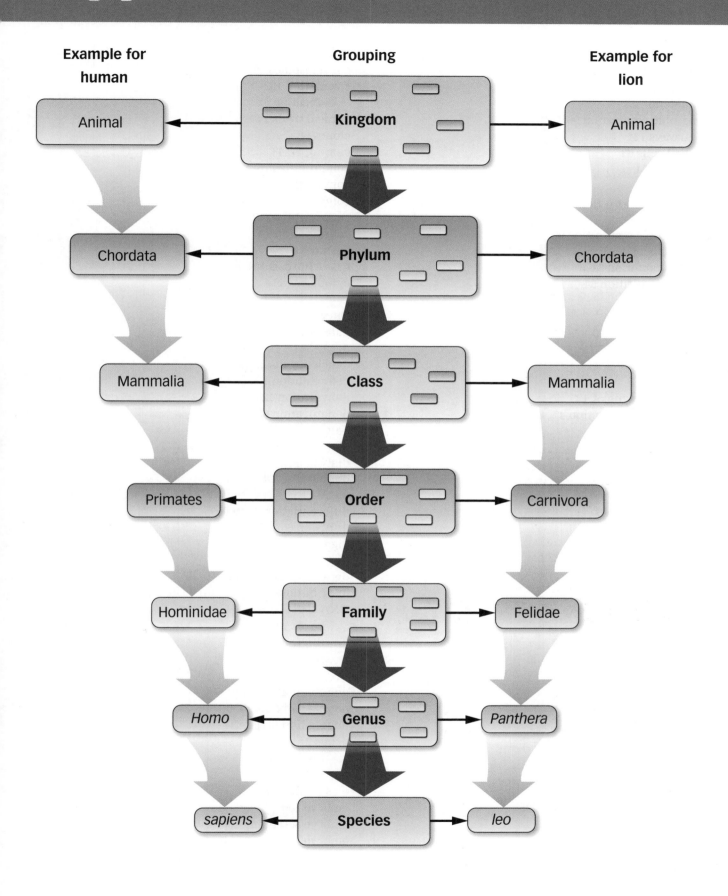

Example for human | **Grouping** | **Example for lion**

Animal ← Kingdom → Animal

Chordata ← Phylum → Chordata

Mammalia ← Class → Mammalia

Primates ← Order → Carnivora

Hominidae ← Family → Felidae

Homo ← Genus → *Panthera*

sapiens ← Species → *leo*

Index

OXFORD
UNIVERSITY PRESS

Great Clarendon Street, Oxford OX2 6DP

Oxford University Press is a department of the University of Oxford.
It furthers the University's objective of excellence in research,
scholarship, and education by publishing worldwide in

Oxford   New York

Auckland   Cape Town   Dar es Salaam   Hong Kong   Karachi
Kuala Lumpur   Madrid   Melbourne   Mexico City   Nairobi
New Delhi   Shanghai   Taipei   Toronto

With offices in
Argentina   Austria   Brazil   Chile   Czech Republic   France   Greece
Guatemala   Hungary   Italy   Japan   Poland   Portugal   Singapore
South Korea   Switzerland   Thailand   Turkey   Ukraine   Vietnam

Oxford is a registered trade mark of Oxford University Press
in the UK and in certain other countries.

British Library Cataloguing in Publication Data

Data available

ISBN 978-0-19-913599-8

10 9 8 7 6 5 4 3 2 1

Printed in Great Britain by Bell and Bain Ltd, Glasgow

Paper used in the production of this book is a natural, recyclable product
made from wood grown in sustainable forests. The manufacturing process
conforms to the environmental regulations of the country of origin.